IDEAS FROM ASTRONOMY
An Addison-Wesley Shortext

Lou Williams Page

ADDISON-WESLEY PUBLISHING COMPANY
Menlo Park, California • Reading, Massachusetts • London • Don Mills, Ontario

About the Author

Mrs. Page received her PhD in earth science from the University of Chicago. She is the author of *Dipper Full of Stars* and *The Earth and Its Story,* and was a member of the writing team of the Earth Science Curriculum Project (ESCP). Together with her husband, Thornton Page, at Wesleyan University, she edited the eight-volume *Sky and Telescope Library of Astronomy,* a series published by Macmillan.

Copyright © 1973 by

Addison-Wesley Publishing Company, Inc.

Philippines Copyright 1973

All rights reserved. No part of this publication may be reproduced, stored in a retrieval system, or transmitted, in any form or by any means, electronic, mechanical, photocopying, recording, or otherwise, without the prior written America. Published simultaneously in Canada.

0-201-05651-8

BCDEFGHIJK-DO-7543

Contents

1 **The Silent Stars** 1
   The Maps of the Sky 2
   Ptolemy 7

2 **The Changing Sky** 11
   The Turning Sky 11
   The Motion of the Sun 15
   The Changing Sky 19

3 **The Moon and the Wanderers** 27
   The Phases of the Moon 27
   Eclipses 29
   The Wanderers 32

4 **A Revolution in Revolutions** 37
   What Does the Turning? 37
   Is the Sun at the Center? 42

5 **Tycho, the Observer** 49
   Reactions to the Copernican Theory 49
   The Nobleman Scientist 50
   Imperfections in the Perfect Heavens 51

6 **One Seventh of a Degree** 59
   The Imperfect Curve 62
   The Laws of Planet Motion 64

7 **Galileo's Telescope** 69
   Galileo's Discovery 69

    A Little Solar System   73
    Resistance to Galileo's Discoveries   77

8 Newton's Explanation   81
    Galileo's Experiment   81
    Apples and Moons   84
    The Effect of Distance   89

9 "But the Earth Does Move!"   91
    Aberration   91
    Stellar Parallax   94
    The Foucault Pendulum   96
    The Military Evidence   98

10 Weighing the Earth, Sun, and Planets   101
    Centers of Mass   101
    Masses in the Solar System   105
    The Planets Attract Each Other   108

11 Uranus, Neptune, and Pluto   111
    Herschel's "Comet"   114
    Perturbations in Uranus' Orbit   117
    Who Cares about Planet X?   118
    And Yet Another   119

12 The Planets   125
    Gravity and the Atmosphere   127
    Long-Distance Chemistry   130
    Dark Lines in the Spectrum   131
    Fraunhofer Lines   133
    Is There Life on Other Planets?   135

13 Other Members of the Solar System   137
    The Minor Planets   138
    Meteor Showers and Comets   142

14 Satellites   149
    Men on the Moon   154
    Artificial Satellites   156

Putting a Satellite into Orbit   157
Satellite Speed and Escape Velocity   158

15   The Sun   161
Sunspots   161
The Chromosphere and Corona   164
The Sun as a Source of Energy   166

16   The Stars as Other Suns   169
The Temperatures of Stars   171
The H–R Diagram   172
Composition of the Stars   175
Pairs of Stars   176
Doppler Shift   177
Masses of the Stars   178
Spectra, Photographs, Radio Waves   179

17   Lives of the Stars   183
Stellar Evolution   184
A Star's Future   189
Baby Stars   191
The Birth of Stars   192
Our Sun's Future   197

18   Our Galaxy, the Milky Way   203
How Big Is the Milky Way Galaxy?   204
The Dust of the Galaxy   205
Is the Sun at the Center?   208
Our Galaxy Must Be Revolving   211

19   Beyond the Milky Way   219
Clouds or Galaxies?   220
The Geography of Galaxies   223
Do Galaxies Evolve?   225

20   The Universe   231
Movement in the Universe   232
The Origin of the Universe   235
Rugs on Curves   237
New Mysteries, New Discoveries   240

Acknowledgements

2-2	Yerkes Observatory	14-2	NASA
3-2	Yerkes Observatory	14-3	Lick Observatory
3-4	The American Museum of Natural History	14-4	NASA
		14-5	NASA
4-1	Yerkes Observatory	14-6	NASA
5-1	The Granger Collection	14-7	NASA
5-2	William Liller	15-1	The Hale Observatories
6-1	Yerkes Observatory	15-2	The Hale Observatories
7-1	Brown Brothers	15-3	Sacramento Peak Observatory
7-3	NASA	15-4	Yerkes Observatory
7-4	Yerkes Observatory	16-4	Yerkes Observatory
7-5	Yerkes Observatory	16-7	Yerkes Observatory
7-6	Yerkes Observatory	16-8	Stanford University News Service
7-7	Lowell Observatory	17-4	Lick Observatory
7-8	Yerkes Observatory	17-6	The Hale Observatories
8-4	Yerkes Observatory	17-7	The Hale Observatories
9-5	top Smithsonian Institution	17-8	Yerkes Observatory
9-5	bottom Museum of Science, Boston	17-9	Lick Observatory
		17-10	Lick Observatory
10-6	Yerkes Observatory	17-11	The Hale Observatories
11-1	Lick Observatory	17-14	Yerkes Observatory
11-2	The Royal Society of London	18-2	Yerkes Observatory
11-3	Yerkes Observatory	18-3	The Hale Observatories
11-5	Lick Observatory	18-4	Yerkes Observatory
11-7	Lick Observatory	18-5	Lick Observatory
11-8	Yerkes Observatory	18-8	Peter van de Kamp, Sproul Observatory
12-1	right Yerkes Observatory		
12-1	left NASA	19-1	Hans Pfleumer
12-2	Yerkes Observatory	19-2	Yerkes Observatory
12-7	The Hale Observatories	19-3	The Hale Observatories
13-1	The Hale Observatories	19-4	The Hale Observatories
13-2	Yerkes Observatory	19-5	The Hale Observatories
13-3	Yerkes Observatory	19-6	The Hale Observatories
13-5	The Hale Observatories	19-8	left The Hale Observatories
13-6	Authenticated News International	20-1	Lick Observatory
		20-2	The Hale Observatories
14-1	Lick Observatory	20-8	The Hale Observatories

1

The Silent Stars

If it's clear tonight, go out and look at the sky. If there is no moon, all you will see are jewel-like points of light, some lying in a misty band across the sky. You could count over two thousand stars on a clear night if you have good eyesight and live in a dry climate, far from city lights. You will see many, many less, however, if it's a moonlit evening or you are in a city.

Above the earth today is the same sky that looked down on the ancient Greeks and Romans. But these people's thoughts about the starry sky were very different from yours. To them, the stars were merely what they appear to be: sparkling decorations which make the dark, bowl-shaped sky beautiful and somewhat awesome.

While you may appreciate the beauty of this heaven-full of stars as much as they did, your thoughts about it range much farther. You have heard that the stars are huge and hot, that their numbers and their distances apart are so great as to defy the imagination. Questions come to your mind: How do we know that most stars are glowing spheres of gas? That some are old and some are young? Why do we believe that some of these tiny lights are much nearer than others and cool and solid enough to land on? How can we tell which they are? Are there living things on any of them? The Greeks and Romans did not ask these questions, much less answer them.

The stars have remained silent and man has remained isolated from them on the faraway earth. Yet, 2000 years and a million stargazers later, there are answers to these questions and many more. Over the years astronomers used their eyes, aided by instruments which they devised, and reasoned about what they saw.

Each hard-won answer made our picture of the universe more complete. And each new answer raised new questions. The first astronomers wondered what the stars are, and tried to find out. Today astronauts successfully journey into nearby space and astronomers debate the current, hotly disputed theories of the origin of the universe. Let us join this line of investigators and begin to look and reason, starting where the first astronomers began.

The Maps of the Sky

If you face north, you will see the stars shown on the inside back cover of this book. They extend almost from horizon to the *top of the sky.* This place directly overhead is called the *zenith.* When you first look at the stars in the sky, they seem to be randomly scattered. But if you look at them often enough, or hard enough, they will seem to form patterns. These patterns are shown by the lines between stars in the drawing. To the ancients, who spent far more time under the open sky then we do, groups of stars seemed to form the figures of people and animals. These patterns of stars were called *constellations.*

Let us look first at the Big Dipper, a star group that you probably know already. If you have any trouble finding it in the sky, have someone point it out to you. Turn the book so that the drawing of the Big Dipper looks like the Dipper in the sky. Then the other constellations shown will all fall into place between horizon and zenith. See if you can pick out all five of them in the sky.

The Big Dipper is a convenient pointer and yardstick for measurements in the sky. Once you've located the Big Dipper, you can tell which direction is north. Draw an imaginary line between the two pointer stars (the stars in the bowl opposite the handle, as Figure 1-1 shows) to Polaris (the North Star), the end star in the handle of the Little Dipper. As you look at Polaris, you will be facing north. The line from the zenith through Polaris meets the horizon at due north. The pointer stars are 5° apart and Polaris is seven times farther away, or 35°. A distance like this between two stars is measured in degrees. It is the angle between two straight lines from us to the two stars. This angle is part of a circle. A circle is 360°. A right angle is 90°, or one-fourth of the distance all around the sky. The 35° angle from the pointer stars to Polaris is a little more than one-third of a right angle.

THE SILENT STARS / 3

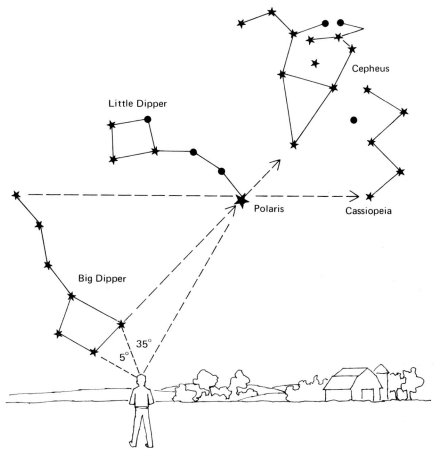

1-1 Can you find Polaris when you look at the night sky?

Cassiopeia (*Kas-ē-ō-pē*-ya) can be found by drawing a line from the end star in the Big Dipper's handle through Polaris. This line will cross the end star in Cassiopeia. As you can see from the diagram, once you find the Big Dipper, the other star groups in this part of the sky should be easier to locate.

The Big Dipper is not really a whole constellation. It is part of the constellation of Ursa Major, the Big Bear, which includes other nearby stars. The Little Dipper is also part of a larger group of stars, the constellation Ursa Minor, the Little Bear. Both bears are rather difficult to make out. For our purposes the clear and plain outlines of two saucepans are more useful. Draco the Dragon, Cepheus (*see*-fē-us), and Cassiopeia are more complete constellations, however. They show how the ancients used the sky as a sort of historical and religious picture book. The stars of Cassiopeia form a clear although shallow *W*, an alphabetical symbol unknown in ancient Greece. But, by including a star just above the middle point of the *W*, they made it into a chair. In fact, they made it into the throne of Queen Cassiopeia of Ethiopea (Figure 1-2), banished forever to the sky by the gods because she boasted too much of her beauty. Her husband, King Cepheus, is there too, looking to us like a country church complete with steeple. But some fanciful stargazers saw him as he looks in Figure 1-2. As constellation drawings go, this is quite a simple clothing of the bare stars! In the Middle Ages, and until about 1850, many well-known artists made imaginative and elaborate drawings of the constellations. Figure 1-3 is part of a sky map made in 1515 by the German artist, Albrecht Dürer. It shows the five star groups you have just seen in the sky, together with others near them.

1-2 (left) The throne of Queen Cassiopeia. Can you see the *W*? (right) The constellation Cepheus looks like a country church to us. Can you imagine it looking like King Cepheus?

Equally elaborate are some of the ancient stories about the constellations and the adventures of the people they represent. Cepheus and Cassiopeia, their daughter Andromeda, son-in-law Perseus,

1-3 If you look carefully at Dürer's sky map above, you will see Draco the Dragon where all the lines intersect. King Cepheus and the throne of Cassiopeia are just to the left of this point. Perseus, their son, is just below the throne to the left. Can you find Pegasus, the horse, and Andromeda?

and even his horse Pegasus are all immortalized as constellations and are the subjects of a 4000-year-old adventure story. Becoming familiar with all the constellations and their legends could keep us busy for many a day and night.

From horizon to zenith, the bowl, or half-sphere, of the sky is studded with constellations. Even the earliest astronomers guessed and probably believed that there were stars below their horizon. And, of course you know that wherever you are on earth, there are stars in the sky. The sky, then, isn't just the half-sphere that we see. It must be a complete sphere encircling the earth. The early astronomers thought of it as a hollow sphere of clear crystal on which the stars were pasted like jewels. They called it the *celestial sphere,* shown in Figure 1-4.

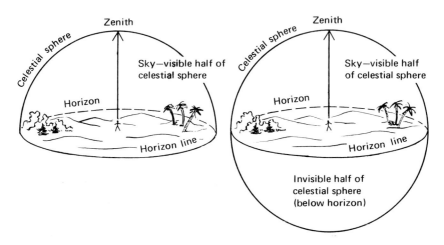

1-4 (left) The sky we see looks like the half-sphere. (right) The sky is really a full sphere.

If you could speak tonight by telephone with others looking at the sky at the same time from many widely scattered places on the earth, you would find that each person had his zenith at a different place among the stars, as in Figure 1-5. Therefore, the horizon line of each would also lie at a different place among the stars—it would cut the celestial sphere along a different line. The visible half of the celestial sphere would not be identical for any two of these people, although two fairly near each other—one in New York City and one in Philadelphia, for instance—would see almost the same stars. An observer at the North Pole would see the Dippers, Draco, Cassiopeia, and Cepheus surrounding his zenith. Observers in Canada and the United States would see these constellations successively lower in the sky. At the equator some of these constellations are below the horizon. As you travel southward more of them are hidden from view, and other stars, never seen in the northern hemisphere, appear.

By putting together observations from all over the earth, the individual stars and the constellations can be drawn on the celestial

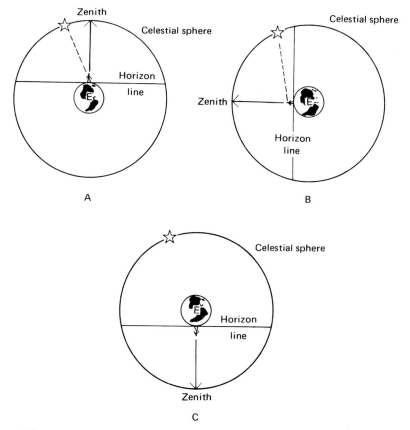

1-5 A person's zenith and horizon lines change depending on where he is on earth. While the star in the diagram above is near *A*'s zenith, it is near *B*'s horizon. *C*, of course, cannot see the star.

sphere—just as the oceans and continents, mountains and rivers, cities and towns can be plotted on a globe of the earth. The stars and constellations thus become *skymarks* in the same way that features on the earth are landmarks.

Ptolemy

About 150 A.D., an astronomer named Claudius Ptolemy (*tall*-eh-mē) lived in Alexandria, Egypt. Ptolemy reached many conclusions, like those you are making, from many years of observing the sky. He recorded these conclusions in a book called *The Almagest*. Ptolemy concluded that the earth was at the center of the sphere of the stars, and very small—a mere dot—in com-

8 / ASTRONOMY

parison to that outer sphere. If the earth were not at the center, he pointed out, the horizon would divide the celestial sphere into two unequal parts. In Figure 1-5 the earth and the starry sphere had to be drawn about the same size in order to fit the drawing on the page. And, as you can see, the horizon in Figure 1-5 does divide the celestial sphere unequally. But, as Ptolemy saw, maps of the stars show that the real horizon everywhere divides the celestial sphere exactly in half as in Figure 1-4.

He also noticed that each constellation appears equal and similar from wherever on the earth one views it. If the celestial sphere were not almost infinitely far away from the earth, then the star shown in Figure 1-5 would have to look larger to observer *A* than to observer *B*. Cassiopeia, for instance, would look larger or smaller, depending on whether you were in New York or London. And, if the celestial sphere were not an enormous distance from the earth, the relation of the stars to each other would appear to change, depending on the angle from which you saw them. Yet the *W* of Cassiopeia is the same shape to those who view it near the zenith and to those who see it near the horizon.

From these and other arguments, Ptolemy concluded that the celestial sphere must be very much larger than the earth. In other words, the stars are very bright objects very far away. We see them at such great distances as tiny points of light.

Equator   Chicago   North Pole

1-6 Since the earth is curved, Polaris will be in different parts of the sky depending on our location.

It may surprise you to learn that Ptolemy also concluded that the earth is round. He saw that the earth's surface is curved. As you travel on the sea toward a range of distant mountains, you see their peaks first, then their bases. If the earth were not curved, the whole mountain would come into view at once, when you got near enough to make it out.

Because the stars are almost infinitely far away, people at different places on the earth look in the same direction into space to see the same star. If the earth were flat, straight up would be the same direction for everybody. Then everyone would see the same star overhead at the same moment. But they don't. The zenith of a man in Mexico City is south of the star being seen overhead in St. Louis. And a man in Canada, observing at the same time, has his zenith north of that star. This shows that the direction of the zenith is different at these places. This shows that the earth is curved from north to south.

Test Yourself

1. What would be the best time and place to look at the stars with unaided eyes?
2. What is the name given the point in the sky directly overhead from where you are standing? Would other people see the same star there from other places on earth?
3. Look up *constellation* in a good dictionary and explain the origin of this word.
4. What two main things did Ptolemy conclude about the earth in relation to the celestial sphere?
5. How did Ptolemy know that the stars must be at immense distances from the earth?
6. Would you expect to see the North Star from Sydney, Australia or Rio de Janeiro, Brazil? Explain.

2

The Changing Sky

After you have found the Big Dipper and Cassiopeia in the sky, go out again after 2 or 3 hours to make sure you can still recognize them easily. You will be surprised to see that these constellations are no longer in quite the same place. The Big Dipper is still a dipper and Cassiopeia is still a *W*. You can still draw a straight line from the end star of the Big Dipper's handle, through Polaris, to the end star in Cassiopeia. Polaris is still at the same distance from each end of this line. However, both star groups have swung around relative to the zenith and to the horizon. If you were familiar with the entire sky and knew all the constellations, you would see that all stars change position, even though the constellation shapes and distances from each other remain the same.

The Turning Sky

What happened while you were inside? If you had been watching the sky carefully, the Big Dipper would have seemed to move slowly and continuously. It would have appeared to move counterclockwise (opposite from the hands of a clock) around a specific place in the sky as you were facing north. And, like a clock hand, the Dipper moved with a circular motion. If you made one drawing of it soon after darkness fell and another 6 hours later, your drawings would be as different as those in Figure 2-1. At daylight the Dipper's stars can no longer be seen in the lighted sky. But when darkness falls the next evening, there is the Dipper back in the sky where you first sketched it.

The hour hand on a clock moves around the clock face once every 12 hours. The line joining Polaris with the pointer stars of the Big Dipper is like a clock hand. But this clock hand completes

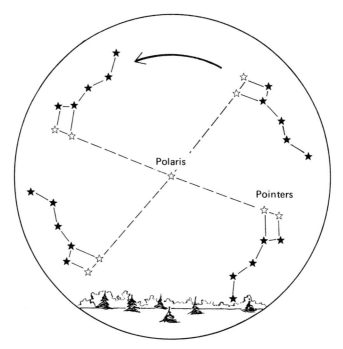

2-1 No matter how the Big Dipper stands in the sky, its pointers always point to Polaris. The broken lines are like the hands of a clock. They revolve around Polaris once every 24 hours.

its circle around a small part of the sky once every 24 hours. The hands of a real clock go around in a circle with their center in the middle of the clock face; they always line up with this center. Carrying the pointer stars with it, the line we have drawn in the sky moves in a circle whose center seems to be the star Polaris. Each of these stars makes a circle around Polaris. The one farther out completes its larger circle in the same amount of time as the nearer stars, which move in smaller circles. Polaris, at the center, doesn't seem to move at all, just like the axle carrying the clock hands or the hub of a turning wheel.

If you were to make more careful measurements, you would see, as the early astronomers did, that even Polaris moves a little. It goes around a tiny circle each 24 hours. The center of this small circle is also the center of the larger circles followed by the stars of the Big Dipper. It is also the center of the even larger circles followed by the stars of Cassiopeia.

What about the other stars in the sky? If you follow their motions through the night, you will see that each one moves in a circle around this same point in the sky. Each one takes 24 hours to complete its circle. Figure 2-2 is a time-exposure photograph of the stars near Polaris. The photograph was made with the shutter of the camera open for 2 hours. The trails show how each star moved during that time.

 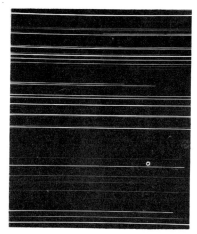

2-2 (left) This is a time-exposure photograph of the stars near the north celestial pole during 2 hours. As you face north, they are moving counterclockwise. The very short curved line near the center was made by Polaris. (right) Time-exposure photograph of stars about 90° from the north celestial pole, near the celestial equator. Can you explain the difference between the two photos?

The place in the sky (which is not a star) around which the stars move is called the *north celestial pole*. By definition, its direction is true north. Stars near the north celestial pole move in circles around it, as do stars farther away. However, the latter appear to be moving east to west across the sky because their circles are so large.

Throughout this rotation, the outlines of the constellations do not change. The stars remain at the same distances from one another, because each one moves around the north celestial pole through the same fraction of a circle in the same length of time. They complete their circles in 24 hours. A circle is 360°. Can you figure out how many degrees a star moves in 1 hour or 1 minute?

If the north celestial pole of your sky were at the zenith, as it is seen from the North Pole of the earth, no stars would rise or set. The same ones would always be above the horizon, continually

14 / ASTRONOMY

circling around the north celestial pole of the sky overhead. However, if you live in the United States, the north celestial pole is not at your zenith. At most, it is halfway up from your northern horizon.

If you watched the stars night after night, you would see that only some of the stars stay above the horizon all around their circles. Stars farther from the north celestial pole dip below the horizon during part of each 24-hour period, as shown in Figure 2-3. Unlike the constellations near Polaris (the Dippers, Draco, Cassiopeia, and Cepheus), those stars rise in the east, move across the sky, and set in the west. The farther they are from the north celestial pole, the shorter the time they are in your sky.

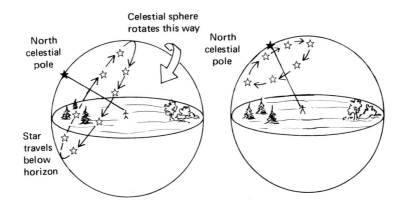

2-3 Some stars seem to move from east to west and below the horizon. Others seem to move in a circle as we watch them.

The ancient astronomers were also watching the stars intently. They had a theory to explain their observations. Ptolemy recorded this theory of the universe in *The Almagest*. The whole celestial sphere, with the stars fixed to its surface, rotates around us once every 24 hours. This rotation carries the stars upward in the east, downward in the west. Its rotation is around an *axis* (a straight line) which passes through the north celestial pole and through the south celestial pole, 180° (a half circle) away, hidden from our view below the southern horizon. If you extended the line in Figure 2-3 from the north celestial pole through the observer to the other side of the celestial spheres, that line would be the axis around which the sky turns westward.

If you consider it carefully, and really observe the motion of the sky, removing from your mind all the things you've merely been *told* are so, you will see that Ptolemy's conclusion was an obvious and reasonable one to draw from the daily motion of the stars. It is pleasant to think of ourselves at the center of the universe, apparently the ones for whom this attractive and changing celestial display was planned. Nothing that you see, or feel, or hear, suggests that the earth has any motion of its own.

The Motion of the Sun

If we are to accept this theory of the universe, it must explain all our observations. What about the sun? How does it fit into the picture? At first glance, this seems simple. The sun—a star much larger than the others—shares the daily westward motion of the celestial sphere. If you watch the sun go down and the stars come out in the darkened sky, you can see which constellation is near the horizon at the place where the sun disappeared. Soon this constellation, carried by the westward rotation of the celestial sphere, follows the setting sun and sinks below the horizon. Then, next morning, just before dawn, you would see another constellation coming up in the eastern sky. As these stars rise higher in the sky, the sun also rises at the same place on the horizon. Then the stars are lost in the sun's brighter light.

Early astronomers reasoned that between sunrise and sunset, while the sun journeys across the sky, the stars are still there. The bright light of the sun blots them out. Ptolemy concluded that as the celestial sphere turns around the earth, it carries the sun and the stars up from the eastern horizon and across the sky. As the sphere turns farther westward, the sun sinks out of sight below the western horizon along with the other stars far from the north celestial pole of the sky.

If you look at the sky both at dawn and at dusk from time to time, however, you will see that different constellations appear with the sun at dawn and at sunset after 2 or 3 weeks. The sun doesn't travel across the sky with the same stars day after day. The early astronomers saw that, unlike the stars, it doesn't stay fixed in place on the turning celestial sphere.

Using the same laboratory and the same equipment that you have—the sky and their eyes—these pioneer scientists made even more observations. They kept track of which stars were near the sun and found that there was a pattern to it: the sun keeps slipping gradually eastward among the stars. In the course of about 365 days it completes one full trip around the sphere of the sky, back to the same place among the stars. This yearly trip is always along the same path (shown in Figure 2-4), crossing twelve constellations which the early astronomers called the *signs of the zodiac. Zodiac*, a Greek word meaning *circle of animals*, is appropriate because some of the constellations of the zodiac represent animals. Each July 10th the sun enters the constellation Cancer;

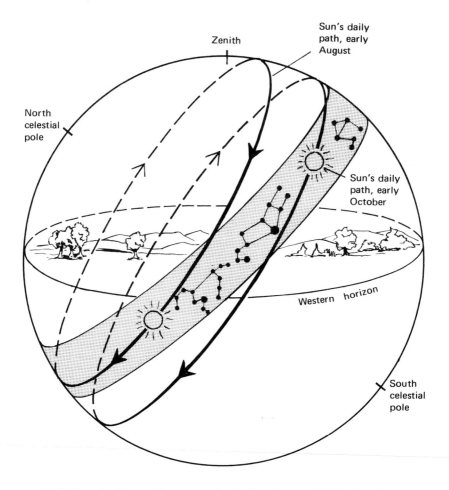

2-4 In early August, the sun sets just before the constellation Leo. In early October, it sets just before Libra. In between, it slipped past the stars of Leo and Virgo along the Zodiac (shaded band) to Libra. (Not drawn to scale.)

THE CHANGING SKY / 17

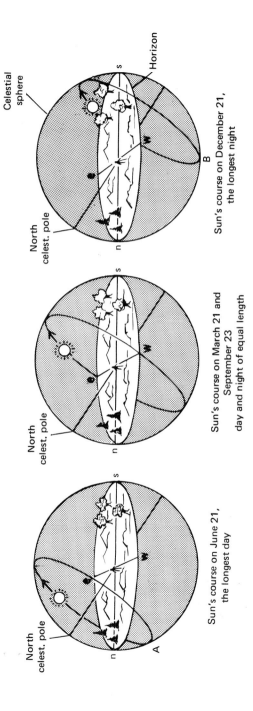

2-5 The sun will move from A to B and back to A in one year's time. The person at the center of the diagrams is in the Northern Hemisphere. For him, the sun is highest in the sky and the days are longest in summer. In the winter when the sun is low in the sky, the days are shorter.

on August 10th, Leo; on September 10th, Virgo; and so on through the year.

Two motions of the sun must be explained: a *daily* westward motion (explained by the turning of the whole celestial sphere) and a *yearly* eastward motion. It seemed obvious to the early astronomers that the sun crawls eastward along a set path called the

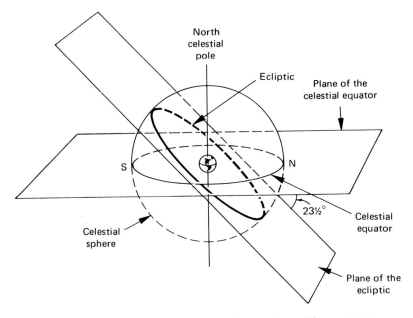

2-6 The celestial equator is in the same plane as the earth's equator, as shown above. However, the ecliptic—the sun's yearly path—is tilted to the equators 23 1/2°. This tilt causes the changing lengths of day and night shown in Figure 2-5.

ecliptic, over the surface of the sphere. In another explanation, the sun was pictured as fastened to a sort of moving ring (still called the ecliptic), much closer to us than the stars on the celestial sphere. In this case, the stars form a background far behind the sun. Hence, the sun need not be bigger or brighter than the stars. The sun's ring turns slowly eastward as the celestial sphere turns westward once a day, carrying the sun's ring (the ecliptic) with it. The sun's ring, however, takes a year for a complete turn in the opposite direction.

A theory should never be more complicated than needed to explain the observations. Either of these theories could explain the sun's yearly eastward motion among the stars. They can also explain the changing seasons and the changing lengths of daytime throughout the year.

When the ecliptic, or yearly path of the sun, is plotted among the stars on the celestial sphere, it is found to be a circle inclined to the celestial equator. The ecliptic is not the middle—the equator—halfway between the north and south celestial poles. In the Northern Hemisphere of the earth, the sun is farthest above the celestial equator (the nearest it gets to the north celestial pole) about June 22. On that day, the sun is visible longer than any other day of the year—as you would expect. When it is below the celestial equator (and the farthest it gets from the north celestial pole), about December 22, it is above the horizon for a much shorter period each day. Between December 22 and June 22, the days gradually become longer and the nights shorter. After June 22, the days gradually become shorter and the nights longer.

The changing lengths of the daily periods of sunlight in a year help bring about the seasons. But more important are the different angles at which the sunlight strikes the earth, according to the sun's position on the celestial sphere at different seasons.

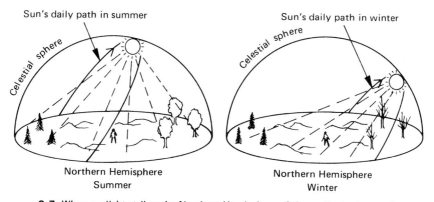

2-7 When sunlight strikes the Northern Hemisphere of the earth at a low angle in winter (right), it is spread out more than when it comes in nearly vertical in summer (left). This change in the angle of sunlight causes the seasons.

The Changing Sky

It is time to go out again and look at the sky. On summer evenings you may have noticed a conspicuous group of three bright stars near the zenith. During the following winter, wishing to show off how well you know the sky (and forgetting the sun's eastward drift among the stars), you might try to point out this star group to one of your friends. Of course you would find that it isn't there, even though you are in the same place at roughly the same time in the evening. Looking more carefully, you notice that although some of the constellations you had seen on summer evenings are shining in the sky, they are in different places with respect to the zenith and the horizon. You also see some new constellations that weren't in the summer sky at the same hour.

What has happened? The early astronomers noticed this change in the evening sky throughout the year. How did they explain it? Like us, these people told time by the sun. Noon, to them as it is to us, is the moment when the sun is highest in the sky that day. It is noon for you when the sun, in its daily journey from east to west, crosses the *meridian*. The meridian is the line on the celestial sphere from the north point of your horizon through the zenith to the south point, shown in Figure 2-8. Since the sun is

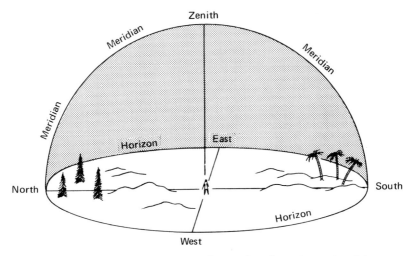

2-8 The meridian is the imaginary line we draw from the north point on the horizon, through the zenith, and down to the south point on the horizon.

slipping backward (eastward) among the stars, 360° per year or about 1° each day, the stars are continually getting ahead of the sun. This is why magazines like *Sky and Telescope* and *Natural History* carry monthly evening sky maps, and why the summer sky holds different constellations than the autumn sky, the winter sky, and the spring sky. On the next few pages (Figure 2-9 a, b, c, and d) you will find star maps that will help you locate stars at four different seasons of the year, soon after dark.

This is, of course, why we had to tell you in Chapter 1 to find the Big Dipper and then twist the book so that the Dipper looked the same as you saw it in the sky. The stars are constantly changing position in the sky. There was no way for us to tell on what date or at what hour you were going to start reading this book.

Test Yourself

1. If you were at the North Pole of the earth, how would you find the star Polaris?
2. What is the *zodiac*? What connection does it have with the sun?
3. What is the *ecliptic*? What is the difference between the ecliptic and the zodiac?
4. What is meant by "your *meridian*"?
5. What is the name of the point in the sky around which the stars, seen from the Northern Hemisphere, seem to revolve?
6. Do you suppose the constellations of the northern sky could be used to tell time, like a clock? If so, how would you use them? Does this clock run fast or slow?
7. Suppose that at 6 P.M. some autumn evening, you saw the Big Dipper low in the northern sky, with its bowl open upward. Later that night you saw the Dipper higher in the sky to your right, with its bowl open toward the left. You saw that the Dipper had rotated 1/4 of the way around its circle. What time is it?
8. Nights are longest around December 22 and days are longest around June 22. About when would you expect night and day to be of equal length? Are you sure your answer is complete?
9. If a friend had been traveling and told you he had been in a place where the *daylight* was longest on December 22, would he be joking or could he have been in such a place?

22 / ASTRONOMY

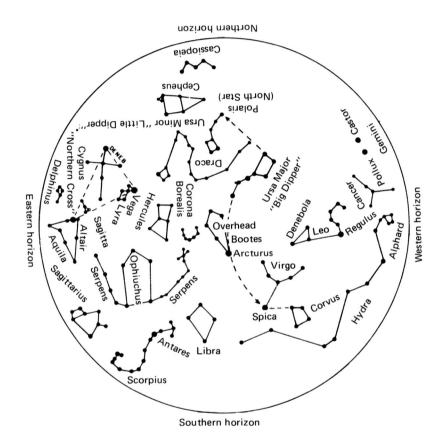

2-9a The evening sky in June.

THE CHANGING SKY / 23

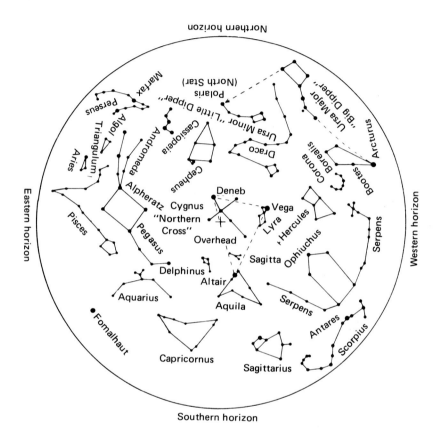

2-9b The evening sky in September.

24 / ASTRONOMY

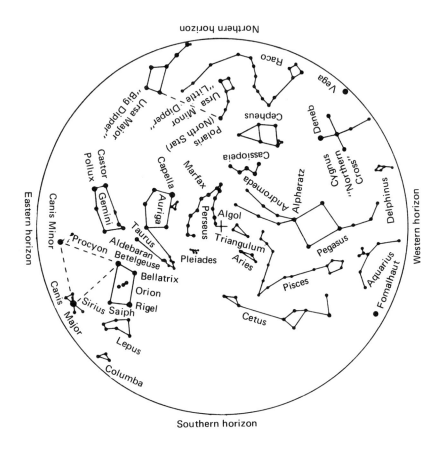

2-9c The evening sky in December.

THE CHANGING SKY / 25

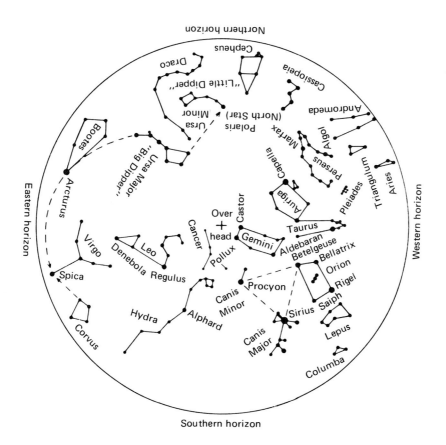

2-9d The evening sky in March.

3

The Moon and the Wanderers

Using a star map in the preceding chapter, you may go out one evening to add another constellation to those you know. Perhaps you find the sky so brightly lit by the full moon that stargazing is impossible. This is just the time to do some moongazing instead. The moon, in the southeastern part of the sky, is steadily moving westward. If you watched it all night, you would see it set toward dawn. Tomorrow night you will see it rise again over the eastern horizon—but almost an hour later. Like the sun, it appears to share in the daily westward turning of the celestial sphere, but it lags a good deal more.

The Phases of the Moon

Watch it for 2 weeks, though, and you will see several differences between the moon and the sun. Most striking of these, of course, are the moon's changes in appearance. You will see it go from full moon, to three-quarters moon (called a gibbous moon), to half-moon, to crescent, *waning* to no moon visible at all. In the next 2 weeks you will see these phases in reverse order, as the moon is *waxing* from crescent to full.

You have found that the moon rises and sets about 50 minutes later each day than it did the day before. Because even the brightest moon doesn't completely drown the light of the stars, it is possible to identify the constellations in which the moon crosses the sky on succeeding nights. When these constellations are recorded, it is evident that the moon is indeed slipping backward (eastward) on a path passing through the twelve constellations of the zodiac.

The moon slips back at a rate of about 12° per day, while the sun slips only 1° per day. As a result, the moon completes a full circle around the celestial sphere once each month. The sun takes a year to complete its circuit. Just as the sun determines the length of the year, so the moon determines the length of the month.

Your records will also show that the *full* moon always rises about sunset and sets about dawn. This means that the full moon is always on the opposite side of the earth from the sun. The thin crescent seen just before and just after the new moon (when it is all dark), rises in the early morning just before the sun or sets in the early evening just after the sun. When the lighted portion of the moon is growing larger (waxing), the moon rises during the daytime hours—you can often see it in the sky with the sun. In its waning phases the moon rises during the nighttime hours. Figure 3-1 shows how early astronomers viewed and explained the moon's phases.

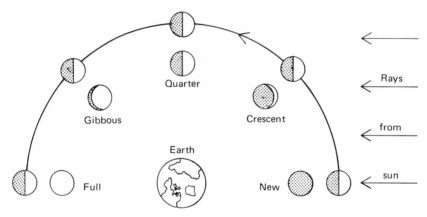

3-1 The moon is shown at four different positions in its orbit around the earth. Half of it is always lighted by the sun's rays. The inner drawings show how the moon looks from the earth.

The positions of the sun and the moon at each moon phase are clues which early astronomers interpreted in the same way we do today. When the moon is on the opposite side of the earth from the sun, the sun lights up the whole surface of the moon facing us; the moon is full. When the moon comes between us and the sun, the far side of the moon that we can't see is lighted, and the dark side faces us. Between these two extremes, as the moon circles eastward around the earth, we see less and less of the moon's lit surface, and then more and more of it.

This explanation points out something else about the moon. Unlike the sun and the stars, it has no light of its own. It shines only by reflected light; it reflects sunlight back to us.

3-2 "The old moon in the new moon's arms." Can you figure out why the dark side of the moon is partially illuminated?

The moon's phases showed the ancients that it could not be way out with the stars on the celestial sphere. If it were, it could not get between us and the sun. It must, they reasoned, be nearer than the sun. They thought of the moon as attached to another moving ring, attached somehow to the sphere, and located inside the ring bearing the sun. The ring carried the moon eastward around the earth once each month, while the rotation of the sphere carried it westward across the sky once each day. It was clear to these observers that the moon must be smaller than the sun. Although the moon appears almost the same size in the sky as the sun, it is nearer to us.

Eclipses

The sun, the moon, and the earth are sometimes lined up with each other. When this happens, the sun or the moon is blotted out in an *eclipse*. If the moon is blotted out, it is a lunar eclipse; if it's the sun, it is a solar eclipse. Lunar eclipses last only an hour or so—solar eclipses much less. Afterwards the sky returns to normal. But these unusual occurrences made most people uneasy. What if the sun didn't come back? Although eclipses of the sun and moon filled their minds with superstitious fear, the early astronomers welcomed them as evidence for their theory. They saw that all solar eclipses take place when the moon is new and all lunar eclipses when it is full moon as Figure 3-3 shows. Of course, if the moon gets directly between us and the sun it can eclipse (cover) the sun, since both appear to be the same size to us. And when the moon and the sun are lined up on opposite sides of the earth, the shadow cast by the earth can darken the moon.

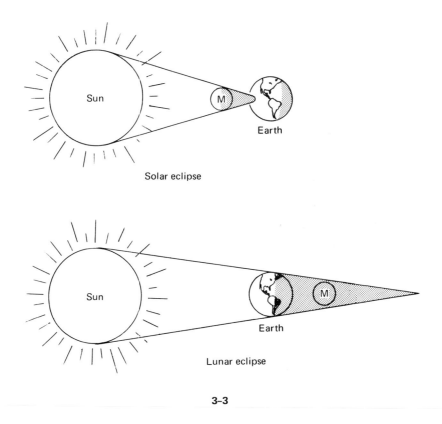

3-3

You may be wondering why there aren't two eclipses each month, one of the sun and one of the moon. There would be if, as the ancients put it, their paths among the stars were identical—if both rings were on the ecliptic. For then the moon, sliding east-

3-4 The sun is seen as a crescent during a partial eclipse.

ward faster than the sun, would have to cover the sun each time it passed by. Careful measurements show, however, that the path of the moon is tipped to the path of the sun (see Figure 3-5). The moon is usually slightly above (north) or slightly below (south) the sun as it passes it. Only where the moon's path crosses the sun's path (as seen against the background of the stars) can there be an eclipse—hence the name ecliptic. It will be a solar eclipse if the sun happens to be in line with the moon and the earth, and a lunar eclipse if the sun happens to be lined up on the opposite side of the sky. The earth's curved shadow on the moon during lunar eclipses also convinced Ptolemy that the earth is round.

A good theory must do more than just explain a few observations. Early astronomers found that knowing the positions of the sun and moon, and the path and speed of each, they could successfully predict when eclipses would occur. It was through such predictions that astronomers became known as wise men.

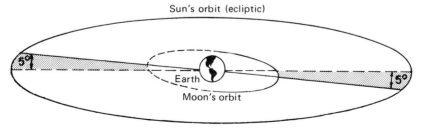

3-5 In the Ptolemaic universe astronomers could predict eclipses. Eclipses only occur when the sun and moon are close to the intersection of the two planes which are inclined 5° to each other.

The Wanderers

As we considered the eastward paths of the sun and moon through the constellations of the zodiac, you may have looked for some of these star groups in the sky. Perhaps you became familiar with Leo the Lion, Figure 3-6 (left), shining almost overhead on springtime evenings. The sickle which forms the forepart of the Lion, with a bright star in the end of its handle, is easy to spot in the sky. Perhaps one summer evening when you went out to see Leo, now stalking along the southwestern horizon, you found the sickle looking like Figure 3-6 (right). That second bright star certainly wasn't there when you first looked at the constellation a month or two earlier. As the summer advanced, Leo set with the sun, and you didn't see him again until early next spring. Then you might have found him with only one bright star again, as in Figure 3-6 (left).

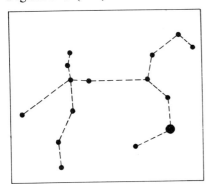 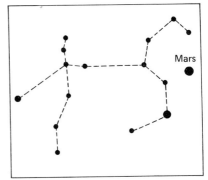

3-6 (left) The constellation of Leo the Lion. (right) The same constellation showing the appearance of the second bright star.

Imagine a group of shepherds, long before Ptolemy's day, whiling away the long nights by watching the stars. This particular night they were looking at the zodiac constellation of Taurus the Bull. It had two bright stars and several fainter ones, forming a letter *V* lying on its side. After a few nights, one of the more careful stargazers noticed that the brightest star of Taurus was no longer in quite the same position relative to the other stars in the constellation. His friends agreed that the constellation did look changed, and that none of the stars in Taurus except this one had moved with reference to the others. Several weeks of observation showed that this bright star in the Bull was slowly moving eastward among the stars. In time, its eastward path took it out of this constellation and into another zodiac constellation, Virgo.

By Ptolemy's time, observations of the sky showed that five "stars" move among the fixed stars. Each one was given a name in honor of one of the gods: Mars (the planet pictured with Leo in Figure 3-6 (right), Jupiter, Saturn, Venus, and Mercury. The Greeks called them *wanderers*—in Greek this word is spelled *planet*.

If you were to chart the motions of Mars, Jupiter, or Saturn over a period of time, you would find that they cross the sky each night from east to west in one of the constellations of the zodiac, and that, in general, they slide slowly eastward among the stars. But every now and then you would see one of these planets slowly make a loop among the stars. After moving eastward for a little more than 2 years (Mars) or a little more than a year (Jupiter or Saturn), it slows up, reverses its path, and proceeds westward for a while before resuming its eastward path. A planet does not repeat its path accurately among the stars as the sun does each year and the moon each month.

The most beautiful of the wanderers is Venus, exceeded in brightness only by the sun and moon. Because Venus is never very far from the sun (east or west), you never see it in the middle of the night. It sometimes shines for nearly 3 hours after sunset in the western sky. Then it is called the *evening star*. Other times it shines for about 3 hours before sunrise in the eastern sky. Then it is called the *morning star*. The very early astronomers had two names for this planet; for when they saw it in the morning sky, they did not realize that it was the same planet they had seen some 6 months earlier in the evening sky. Later they realized that Venus was moving west to east and then east to west on either side of the sun in the sky.

Mercury, very difficult to see, behaves in the same way. About 2 months elapse between its appearance as a morning star and as an evening star.

Ptolemy thought about how these peculiar motions of the planets could be fitted into the theory of the universe. He pictured, as you may now expect, separate bands or rings around the inside of the celestial sphere, one for each planet. Each planet, however, was not fastened directly to its turning ring, as the sun and moon were to theirs. Instead, the planet was attached to a

34 / ASTRONOMY

smaller ring which in turn moved on the main ring, like a wheel rolling on a wheel. The smaller ring or wheel carried the planet around, so that it was sometimes moving in a direction opposite to the movement of the big ring. The rings turned at different speeds. As Ptolemy showed, these movements could produce the looped motions seen in the sky, or the back-and-forth motions of Mercury and Venus.

3-7 (left) A planet may appear to follow a loop as we plot its position among the stars, month after month. (right) Another planet may appear to move from one side of the sun to the other and back again.

By Ptolemy's time, men had been accurately recording and timing the positions of the planets among the stars for several centuries. He estimated the ring sizes and speeds to explain all these observed and recorded motions. This explanation worked very well and was accepted for over a thousand years. The trouble was that the planets didn't move among the stars exactly as the theory predicted. Ptolemy's successors would patch it up by adding more wheels or rings. A planet was supposed to be attached to a wheel whose axle was attached to another wheel, whose axle was attached to still another wheel, whose axle moved on the main ring. Before too long, as many as eighteen wheels moving on each other at varying speeds were needed to explain the past and present positions of one planet in the sky.

Ptolemy's picture of the planets' motions was a mathematical model (see Figure 3-8). He called the smaller rings or wheels *epicycles* but probably never thought of them as material rings or wheels. He used mathematics to build up an explanation of observed movements. It is mathematically possible to represent any movements of the planets if a sufficient number of wheels within wheels is used.

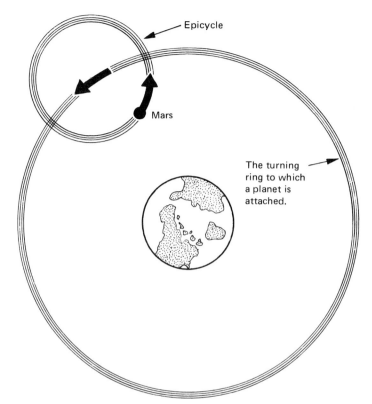

3-8 Ptolemy's conception of the movement of the planets. He drew separate epicycles for each planet to explain its path against the background of the stars.

By 1500 A.D. astronomers were using forty epicycles to explain the past positions of the planets. The theory was becoming terribly complicated. And, worse than that, it could only predict a planet's motion for a few years. It failed to explain all the observations. The reason people accepted it for so long was that they believed that since a circle was the perfect form, motion in a circle was the perfect motion. Because they were dealing with celestial or divine objects, surely the only shapes and movements these objects could possibly have were circular ones. Hence they developed the idea of rings and wheels within wheels. More important, there was nothing to suggest that the solid earth might be moving. Besides, it was unthinkable that earth, the abode of man, could be anywhere but at the center of the universe.

But the difficulties of the Ptolemaic theory set the stage for a questioning of this whole interpretation of the universe. And, as we shall see, a new picture of our universe came from a different interpretation of the movements of the stars, sun, moon, and planets in the sky.

Test Yourself

1. Give a simple definition of an *epicycle*.
2. In this chapter, the *waxing* of the moon is mentioned. What can the moon have to do with wax? Does a good dictionary help you with this?
3. Why are planets called planets? That is, what does this word mean?
4. Why did ancient astronomers have two different names for the planet now called Venus?
5. Since Ptolemy's epicycle "rings" explained all the observed motions of the planets in his time, why was his theory later questioned?
6. Why did people accept Ptolemy's ring or epicycle theory for so long (more than a thousand years) if it became troublesome and less and less able to explain planets' motions?
7. How did Ptolemy interpret an eclipse of the moon?
8. What is meant by *phases* of the moon?
9. Why does a full moon always rise at about sunset?
10. The ancients thought that the moon revolved around the earth, and so do we. But if this is true, why do you suppose we always see the same face or side of the moon, and never its "back" side?
11. How did the ancients know that the moon must be smaller than the sun?
12. What has an eclipse to do with the ecliptic?

4

A Revolution in Revolutions

As far back as 300 B.C., Aristarchus (Ar-i-*star*-kas), a Greek philosopher, had pictured the sun at the center of the universe, with the earth moving around it. His idea did not cause much of a ripple in the scientific world. It seemed obvious that the earth wasn't moving. Also, the problem of accurately predicting the planets' paths or explaining thousands of careful observations of their positions in the sky had not yet arisen. An earth-centered universe was obvious, satisfying, and adequate. So, until 450 years ago, the theory of Ptolemy held sway. For the final 1400 years of its reign—and especially throughout the Dark Ages—astronomers were kept busy patching up the epicycles to agree with the observations. There were, by then, forty of these wheels-within-wheels, each moving at a different speed. This made it a difficult job to predict where a planet would be in the sky on a given date. And even after all this tedious work, the predictions were often not too accurate.

What Does the Turning?

In 1543, the theory of a Polish monk was published under the title, *First Account of the Revolutions of Nicholas Copernicus*. The revolutions were those of the earth and the other planets around the sun. But as we think about this title today, it has a second meaning. The book began a revolution in ideas about the universe.

Copernicus, like astronomers before him, believed that the stars are fixed to a celestial sphere and that the planets and the moon move in circles, each at unchanging speed. But, he said, the celestial sphere is motionless. The earth is rotating daily. The sun is motionless too; the planets revolve around it and not around the earth. In his theory, the earth is far from being the central body of the universe. Instead, it is a planet, like the five wanderers seen only as small lights in the night sky. This demotion was hard to swallow. It was even harder to believe that the solid earth could be in motion. What evidence did he give to back up his ideas?

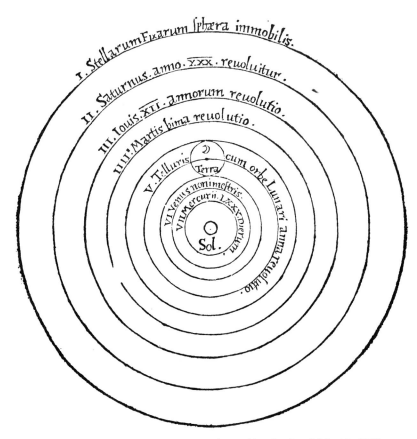

4-1 The Copernican solar system, as pictured in a book published in 1566.

First of all, Copernicus pointed out that a person moving at a constant speed may not be aware that he is moving if this motion requires no effort on his part. Have you ever looked from the window of a moving train and felt that the train was standing still while everything outside it was moving rapidly backward? Or, when a car parked parallel to yours backed out, have you ever felt that your car was slipping forward? Motion of the observer and the observed cannot always be distinguished.

In the same way, Copernicus said, the daily westward motion of the celestial sphere, which appears to carry the stars, the sun, the moon, and the planets from east to west each day, is only an apparent motion. In reality, he said, the earth itself is rotating eastward, one complete rotation each day, while the stars remain motionless. The earth's rotation is around an axis (a straight line) passing through its North Pole and its South Pole. One end of this axis points toward the north celestial pole and the other toward the south celestial pole.

Copernicus also said that the apparent annual motion of the sun was due to an annual motion of the earth in a circular orbit around the sun. As the earth moves in this orbit, we see the sun against a changing background of constellations.

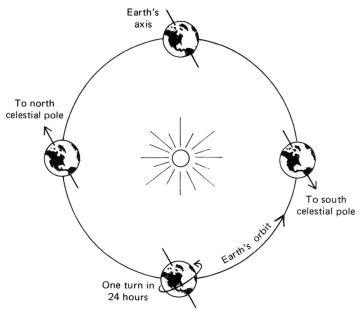

4-2 Copernicus suggested that the earth moves around the sun once each year and rotates eastward on its axis once each day.

Study Figure 4-3. Imagine a huge sheet of paper which passes through the earth at its equator (*B*) and continues outward to cut the celestial sphere at its equator (from *A* to *C*). This is the *plane* of the equator. The ecliptic (see Figure 2-5, page 17), as we have

40 / ASTRONOMY

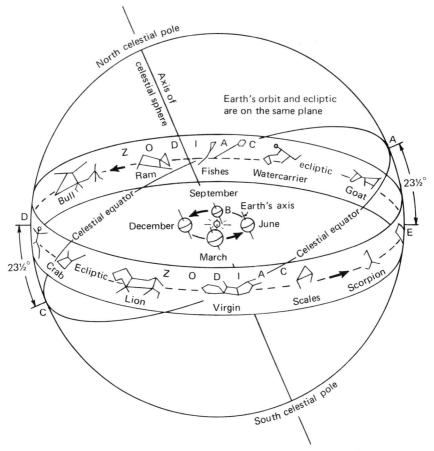

4-3 A Copernican view of the universe. (Not drawn to scale.)

seen, is the observed yearly path of the sun against the background of the stars. Now imagine a second sheet of paper across the hollow interior of the celestial sphere, cutting it along the ecliptic (from *D* to *E*). This is the plane of the ecliptic. The plane of the ecliptic is tilted 23½° to the plane of the celestial equator.

To Ptolemy, the sun's path around the earth lay on the plane of the ecliptic. Figure 2-5 shows the sun at three positions on its path (June 21, March 21, and September 23, and December 21). The sun, traveling around the earth, passes north of the earth's equator and then south of it. Copernicus, on the other hand, saw the system as in Figure 4-3. To him, the plane of the ecliptic contained the path of the earth around the sun. As the earth travels around the sun, the earth's equator is first above the sun and then below it. Notice that the earth's equator is *always* tilted the same amount (23½°) to the plane of this yearly path.

Now study Figure 4-4. From this diagram you can see why winter is colder than summer and why the Northern Hemisphere is shivering while the Southern Hemisphere is enjoying summer. And you may be surprised to know that Ptolemy explained the seasons as well as Copernicus did. In December, the Northern Hemisphere is having its shortest periods of daylight and the Southern Hemisphere its longest. In June these conditions are reversed. About December 21, the sun is lowest in the sky in the Northern Hemisphere and highest for the Southern. Copernicus said that the varying periods of daylight were due to the tilt of the earth's axis. Ptolemy said that it was due to the tilt in the sun's path.

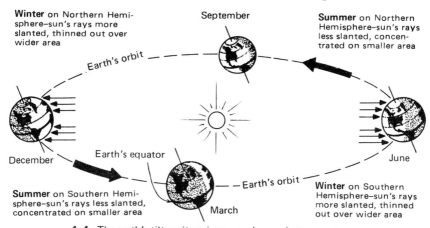

4-4 The earth's tilt on its axis causes changes in temperature.

In both theories it is the changing angle of the sun's rays as they strike the earth (Figure 2-7, page 19) that is the chief cause of the seasonal changes in temperature. Both theories explain equally well why the days and nights are of equal length at the equator, and why the north and south polar regions each have 6 months of darkness and then 6 months of continual daylight. In both theories, on about March 21 and September 23, the ecliptic crosses the equator, and so both explain why the days and nights are of equal length.

The observed seasonal differences in temperature could be produced by the movement of the sun along a path inclined to the earth's equator, together with a daily rotation of the celestial sphere as Ptolemy believed. Or they could be produced, as Copernicus suggested, by the movement of the earth along a path inclined to its equator, together with a daily rotation of the earth on its axis.

Is the Sun at the Center?

So far, the theory of Copernicus does not seem to help much. It substituted movements of the earth—hard to believe—for motions of the celestial sphere and the sun. It told no more and no less about the causes of the seasons. And, as you probably have guessed already, it left the moon moving around the earth, for there is no other way to explain its phases.

However, if the earth is included among the planets, and all six of them move around the sun in circles, the observed motions of the planets are more easily explained. In the sky, Mercury and Venus move from one side of the sun to the other. Venus never gets very far from the sun; Mercury even less. Copernicus said that this is because they move in smaller orbits closer to the sun than the earth's orbit.

Copernicus then explained the looping paths of Mars, Jupiter, and Saturn by saying that they move around the sun in larger orbits, farther from the sun than the earth's orbit. As you may remember from Chapter 3, a planet farther from the sun will appear to follow a loop at regular intervals. Although generally moving eastward with respect to the starry background, it will appear to stop, move westward for a while, and then resume its eastward course along the celestial sphere each time the earth passes it.

Copernicus went on to explain that the looping motion of planets in the sky takes place when the earth overtakes and passes one of these planets. This passing occurs at regular intervals, different for each planet. For Jupiter it is every 1.09 years. This is not the time that Jupiter takes to complete its circle around the sun (Jupiter's *period*). Nor is it earth's period (1 year). However, from these two bits of information Copernicus was able to find the period of Jupiter, the time it takes Jupiter to go around the sun.

We can see how he did it by an analogy. Consider the two hands of a clock. These move around a spot in the center of the clock's face. The minute hand has a period of 1 hour; the hour hand has a period of 12 hours. Thus, at twelve o'clock the two

A REVOLUTION IN REVOLUTIONS / 43

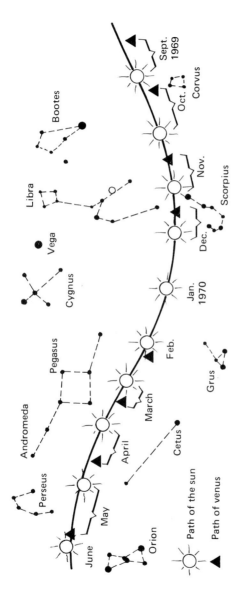

4-5 The path of Venus and the sun from September, 1969 to June, 1970. During September Venus rose about 2 1/2 hours ahead of the sun. During October it rose shortly before dawn. By November it had moved closer to the sun in the sky—it was still a morning star, but visible for a much shorter period before dawn. By January, 1970 it rose less than half an hour before the sun. Venus was not visible during most of January. During February it appeared again, now as an evening star, setting half an hour after the sun did. During March and April it was prominent in the evening sky—in April setting about 1 1/2 hours after the sun; by June 1 it set over 2 hours later than the sun.

44 / ASTRONOMY

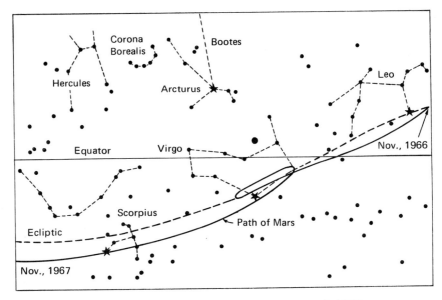

4-6 The path of Mars (November 1, 1966 to November 1, 1967) moves from right to left in the above diagram.

hands lie along the same line to the center axle of the clock face. By one o'clock the minute hand has completed one revolution; it is again pointing to twelve. But it does not line up with the hour hand, because the hour hand has moved along 1/12 of its circle in that hour. It is not until about 1:05 that the two hands line up again. Again the minute hand completes a revolution, but by then the hour hand has moved on, and they line up again at about 2:10. The interval between the lining up is about 1 hour and 5 minutes (1.09 hours).

If we compare the minute hand (period = 1 hour) to the earth (period = 1 year), and the interval between the lining up of the clock hands (1.09 hours) to the interval between Jupiter's loops, when the two planets are lined up, (1.09 years), then Jupiter's period may be compared to the period of the hour hand (12 hours). So it should not surprise you that Copernicus' calculations showed Jupiter's period to be about 12 years. The period of Mars and Saturn may be worked out by similar reasoning because the intervals between the loops of these planets can be timed. Copernicus showed that Mars' period is a little over 2 years, and Saturn's about 30 years. The relative distances of the five wan-

Table 1. Average distances of planets from the sun measured in AU's (astronomical units).

Planet	Determined by Copernicus	Modern determinations
Mercury	0.36	0.387
Venus	0.72	0.723
Earth	1.00	1.00
Mars	1.5	1.52
Jupiter	5	5.20
Saturn	9	9.54

derers are shown in Table 1. Figure 4–7 shows the orbits around the sun that he drew. Since the distance of the earth from the sun was not known, he called the distance 1. Today it is called 1 *astronomical unit* or 1 *AU*. The other planets' distances from the sun are based on this measurement.

By comparing the periods of the planets with the relative sizes of their orbits, he found out that the nearer a planet's orbit is to

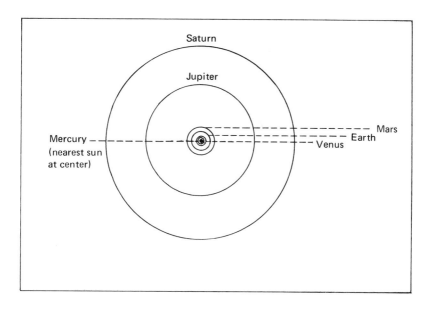

4-7 Copernicus believed that the planets orbit around the sun and that the earth is not the center of the universe.

the sun, the faster the planet moves along it. For instance, while the earth moves a mile, Mars moves only about 3/4 of a mile and Jupiter less than 1/2 a mile.

In general, Copernicus' system could explain the observed motions of the *solar system*—sun, moon, and planets. However, in detail it was a little off the mark. Some of his predictions of planet positions were not quite accurate—perhaps one degree or so off. To make them fit precisely, he had to add forty-eight epicycles to his system—eight more than the Ptolemaic system had at that time. Then the observations fit. The chief advantage over Ptolemy's system was Copernicus' explanation of why the eastward paths or Mars, Jupiter, and Saturn in the sky were interrupted by westward, looping motions, whereas Mercury and Venus seemed merely to move from one side of the sun to the other. It also explained why these loops of Mars, Jupiter, and Saturn always take place when the planet is opposite from the sun, with the earth between.

Test Yourself

1. Copernicus said that the earth and all the planets moved in circles around the sun, but, like Ptolemy, he said the moon moved around the earth. Why did he think so?
2. What is the *period* of a planet?
3. What is an *astronomical unit*?
4. The planets travel eastward in their paths through the sky, but from time to time some of the planets will appear to slip westward, in effect to move backwards, for a while. Why, according to Copernicus, does this happen?
5. How do planets nearer the sun than the earth is appear to move through the sky, as seen from the earth?
6. How well or completely did Copernicus prove that the earth moves around the sun?
7. Copernicus' theory did not catch on right away, but it had one great practical advantage over Ptolemy's. What was the advantage? (It was not just "simplicity.")

8. How does it happen that the axis of rotation of the earth, which passes through its north and south poles, also happens to pass *exactly* through the north celestial pole and the south celestial pole? Think about this. The answer is simpler than it may look to you at the moment.
9. What do you suppose led Copernicus to work out a new theory of the universe, that denied the earth its "common sense" position at the center of things?
10. The ancients thought that the moon revolved around the earth, and so do we. But if this is true, why do you suppose it is that we always see the same face or side of the moon, and never its "back" side?

5

Tycho, the Observer

Copernicus didn't prove that the earth rotates and revolves around the sun. But he firmly believed that it did, and astronomers had to agree that his theory explained observed movements in the sky just as well as Ptolemy's accepted theory. Indeed, astronomers soon found that tables of the future positions of the planets, based on Copernicus' theory, were usually as accurate as those calculated by the Ptolemaic method.

Reactions to the Copernican Theory

It took a few decades for people to catch on. Then Copernicus' theory caused a tremendous stir. Almost everyone was against it. It seemed an affront to common sense. Even today, when you know that the earth rotates and revolves, it is hard to imagine it whirling like a top as it rushes through space. If the earth were rotating, people asked, wouldn't everything near the equator go flying off? Copernicus had foreseen that objection. His rebuttal was that the equator on the larger celestial sphere would have to be going even faster than the earth's so that the stars could get all the way around their large circles each day. (Just as a point on the rim of a wheel moves through a greater distance each minute than the hub does.) He pointed out that there would be an even greater danger of the celestial sphere flying apart.

The scholars said that this was nonsense. Hadn't Aristotle, most revered of the Greek philosophers, said that the celestial objects— stars, sun, moon, and planets—were quite different from earthly objects? Hadn't he clearly stated that natural laws, which apply to earth and the things on it, do not apply to the divine objects? Of course the celestial sphere could rotate as fast as it wanted to, undisturbed! But the earth couldn't. Besides, it was like social climbing to class the dirty old earth as one of the divine celestial objects.

Some of the clergy liked Copernicus' theory even less, but for a different reason. It was wicked, they said, to think that the abode of man, obviously favored by heaven, might not be the center of the universe. It was sinful to think of it disgracefully chasing around the sun, in the company of the wanderers.

Many astronomers also opposed the new theory, although most agreed that it did explain (for the first time) why the outer planets have looping paths in the sky and the inner ones don't. Many of them used it in determining the future positions of the planets. Such predictions were important to them because, at that time, *astrology* (the telling of fortunes by the positions of the sun, moon, and planets in the zodiac) was tremendously popular. Almost everyone who could afford it consulted an astrologer, and many a king and nobleman hired one to work for him alone. The astrologer had to know the exact position in the zodiac of each planet at future dates and hours so that he could advise his employer when to wage wars, sign treaties, or be especially wary of enemies. Many of these astrologers were really astronomers, who used the equipment and leisure provided by their employers to carry on astronomical studies.

The Nobleman Scientist

Among the astronomers who did not accept the Copernican theory was a Danish nobleman named Tycho Brahe (*Tee*-Ko *Brah-he*), born in 1546, 3 years after Copernicus' death. He was employed by the King of Denmark, who built him an elaborate astronomical observatory in a comfortable island castle. The many instruments in this observatory were more precise than any used before by astronomers. Tycho designed most of them and supervised their construction. But he did not have a telescope; this invention came after his time.

For 20 years Tycho spent every clear night observing the sky. He was the most skilled observational astronomer that the world had seen, and one of the most famous of all time. His observations of the positions of the stars in the sky—all made without a telescope—were very accurate.

5-1 Tycho Brahe.

Imperfections in the Perfect Heavens

In 1572 Tycho saw a new star appear in the constellation Cassiopeia. It grew in brightness until it rivaled Venus. He watched it for 16 months until, after gradually becoming dimmer, it finally disappeared. His most careful measurements showed not the slightest change in its position relative to the other stars. Therefore he knew that it belonged on the celestial sphere. His faith in Aristotle, who had said that the heavens were perfect and unchangeable, was shaken.

Five years later, a brilliant comet like the one in Figure 5-2 appeared in the sky for several weeks. Aristotle had taught that

52 / ASTRONOMY

5-2 Photograph of a 1965 comet, Comet Ikeya-Seki, photographed by William Liller on October 29. The lights of Pasadena, California can be seen in the foreground of the picture.

comets were something abnormal in the earth's atmosphere. Tycho, however, found that the comet was not a small object near the earth, but a large object much farther away than the moon. He reached this conclusion by trying to measure the comet's parallax.

Parallax is the change in position of an object against a more distant background as an observer moves sideways. It is the difference in direction of the line of sight to an object from two different places. Hold a pencil at arm's length. Close your left eye and look at it. Then look at it with only the left eye. The pencil will appear to have shifted sideways to a different place against the background of the room. Do this with the pencil much closer to your face. The movement of the pencil will seem greater. If you focus on something very far away, such as a distant tree, and close first one eye and then the other, you won't see any shift at all.

TYCHO, THE OBSERVER / 53

Study Figure 5–3. If you stand at location *A*, in the left diagram, and view a chair, it will seem to be in front of the picture on the wall. When you move to location *B* and look at the chair, you see it to the left of the picture. The line of sight to the chair has shifted by an amount equal to angle *X*, called the parallax angle. Now, if you move the chair closer, as in the right-hand diagram, and change from position *A* to position *B*, the lines of sight through the chair make a larger parallax angle *Y*.

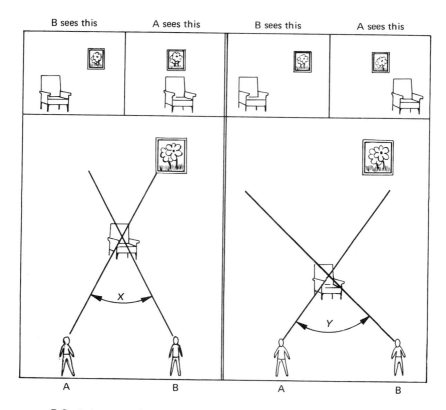

5-3 A demonstration of parallax. Notice that if an object (a chair in this case) is closer to the observer, the angle made by the lines of sight (of two observers the same distance apart) through the object becomes larger. What would happen to the angle if the object were very far away?

The distance between the two observers (or the two places of observation) can be measured. Knowing this distance and determining the parallax angle with special equipment, the distance to the object can be determined mathematically.

If Tycho could have compared his line of sight to the comet with those of other observers at the same distance away, he could have easily determined how far away the comet was—whether it was part of the earth's atmosphere or way out in space. But with the poor communication of his time, and isolated on his island in northern Europe, he couldn't do this. He wanted to find out if it was much nearer or much farther than the moon, and he did this in a most ingenious way.

He did it in terms of a rotating celestial sphere. Today we describe it in terms of a rotating earth. While Tycho was observing, his island had been carried eastward 500 miles each hour on the rotating earth. So, 10 hours later he was observing the moon from a different place in space, as Figure 5-4 shows.

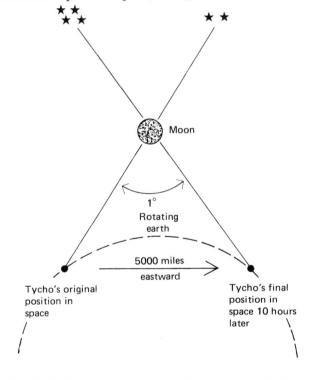

5-4 If the earth were not rotating, and Tycho were still in the same position in space at the end of 10 hours, he would have seen the moon (after its 5 1/2° of movement among the stars) against the background of stars ★★. But the rotation of the earth carried him eastward so that his line of sight then placed the moon 1° west of stars ★★. (The angle in the diagram of course could not be drawn to scale.) This made it look as though the moon had slipped back only 4 1/2°. Yet the next night the moon was just where it should have been if it had been slipping back 5 1/2° each 10 hours.

He measured the parallax of the moon—it amounted to 1° each 10 hours. But meanwhile, the comet appeared to have no parallax. It was too small for him to measure. This meant that the comet was farther away than the moon. With his instruments he could measure lines of sight between stars down to almost 1/60°. So, because he couldn't find *any* parallax, he concluded that it must be less than sixty. If the comet were three times farther than the moon, its parallax would have been about 1/60°. He was sure that the comet was at least sixty times farther than the moon, and probably as far from the earth as Mercury. This would mean that its orbit crossed that of Venus.

Tycho's nightly observations showed him that the comet appeared to move first east and later west of the sun. Tycho reasoned that the comet moved around the sun and that its path came close to the sun. His data showed him that the comet did not move at a uniform speed. He also observed six other comets during his lifetime. He calculated their orbits (as Copernicus had for the planets) and found that they were, in his words, "not exquisitely circular, but somewhat oblong."

Through his observations on comets, Tycho demolished four revered principles of the Aristotle–Ptolemy camp.

(1) The planets couldn't be carried on crystalline spheres, as Aristotle thought, or even on solid rings and wheels. If they did, comets could not move freely through the orbits of these planets. They would bump into the spheres, rings, or wheels.

(2) Not everything moves around the earth—comets move around the sun.

(3) Celestial motions are not all circular—comets move in elongated orbits.

(4) One type of celestial object, a comet, does not move with uniform speed in its orbit.

Nevertheless, Tycho did not accept the Copernican system. For him, the chief stumbling block was the same bit of evidence that made Ptolemy sure that the earth isn't moving: no parallax of the stars. Each constellation appears equal and similar not only from wherever you view it on earth but whenever you view it during the year. If the earth were moving in a yearly orbit around the sun, both Tycho and Ptolemy believed that parallax would change the outlines of the constellations during the year. Ptolemy, you will recall, argued that the always identical appearance of the constellations from everywhere on earth indicated that "the earth is but a point in space," compared to the celestial sphere. It is odd that neither he nor Tycho went one step farther to conclude that the sphere of the stars is so big that even the earth's movement in an orbit couldn't produce a parallax large enough for them to measure.

Hoping to find *stellar parallax* (parallax of individual stars), Tycho measured the positions of a thousand different stars over and over again, with painstaking care. Because he found no parallax, he concluded that the earth must be fixed. But he compromised between the Ptolemaic and Copernican systems as Figure 5-5 shows. He considered the earth at the center, the sun revolving around it each year, and the other planets traveling in circles around the sun in the same orbits that Copernicus gave them. Since, as we have seen, the observed motions of celestial bodies in the sky are explained equally well by revolution of the earth around the sun or by revolution of the sun around the earth, this theory explained the observations as well as the Copernican theory did.

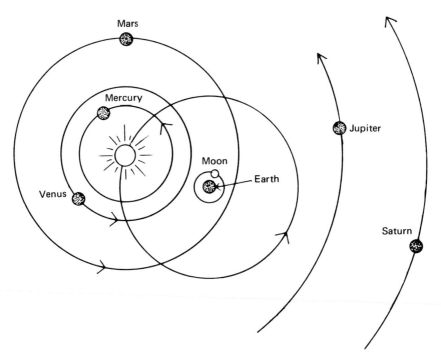

5-5 Tycho's theory of the movement of the sun and planets.

TYCHO, THE OBSERVER / 57

Copernicus' interpretation

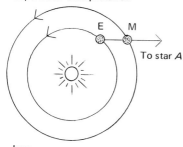

Tycho's interpretation

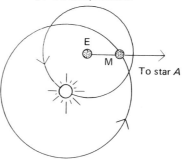

June
Earth, Mars, star A lined up;
Mars moving eastward among the stars.

Copernicus' interpretation
(sun is fixed)

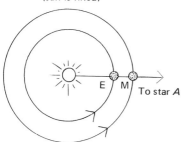

Tycho's interpretation (earth is fixed)

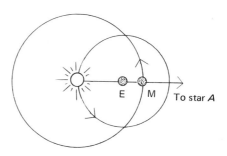

April
Sun, earth, Mars, star A lined up;
Mars in mid-loop
in the sky.

Copernicus' interpretation

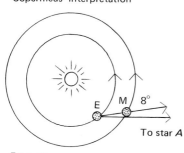

Tycho's interpretation

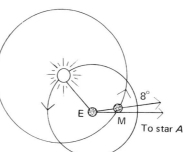

February
Line earth–Mars makes 8° angle with line from earth to star A.

5-6 Copernicus' and Tycho's explanations of the movement of Mars from data taken in February, April, and June.

Tycho's theory was more complicated and troublesome. However, because he had the earth as the center of the universe, he did not tread on so many toes as did Copernicus. People listened and became more willing to consider interpretations of the universe other than Ptolemy's. Figure 5-6 shows the differences in the two men's theories.

Test Yourself

1. Can you think of an important difference between an astronomer and an astrologer?
2. What is *parallax*?
3. What first shook Tycho's faith in Aristotle, who had pronounced the heavens perfect and unchangeable?
4. What discoveries enabled Tycho to demolish four important principles that had been laid down by Aristotle and Ptolemy?
5. Why did Tycho refuse to accept Copernicus' theory, although he accepted many of its features?
6. In what main way did the theory that Tycho worked out differ from that of Copernicus?
7. Since two observing points are needed for measuring parallax, and Tycho had only one (his castle), how did he manage? Why do you suppose his method worked for the moon? Would it have been easier if the moon were nearer the earth?
8. Are we smarter or less prejudiced than people of Tycho's or Copernicus' time, since we do not insist that the earth is the center of the universe? Discuss.
9. Why do you suppose that Tycho looked for stellar parallax by measurements of the same star 6 months apart, rather than by measurements at dusk and again at dawn?

One Seventh of a Degree

Tycho left Denmark in 1597, shortly after a new king ascended the throne. Tycho had been harsh to the people on his island and had ignored the King's warnings to curb his abuse of power. The new king trimmed Tycho's huge allowance to a more modest, but still adequate amount. Tycho left Denmark in a huff and became Imperial Mathematician of Bohemia, at Prague, in what is now Czechoslovakia. There he spent the few remaining years of his life studying the data accumulated in his 20 years of observation.

In 1600, the year before Tycho's death, a young mathematician named Johannes Kepler became his assistant. Kepler, a former high school teacher in Austria, had become famous as an astrologer and prophet with his *Almanac* for 1595. He later succeeded Tycho as Imperial Mathematician and spent the next 25 years altering the Copernican theory to agree with Tycho's observations—much more exact and far more closely spaced than those available to Copernicus. Tycho had told his assistant to use his newer theory, but Kepler wisely chose to work with the Copernican system. As we saw in Figure 5-7, the two theories explain motions in the sky equally well, but Copernicus' system is simpler than Tycho's. The task would be long and hard enough; no need to complicate it further.

Tycho was pleased to have the young Kepler join him as an assistant. Tycho was growing old and he hoped that Kepler would complete his life's work. He first assigned Kepler the task of plotting the orbit of Mars, the planet whose orbit was best observed and seemed trickiest. Kepler found that Mars' position in the sky was 5° away from where the Copernican theory (now 60 years old) predicted it would be. During the first few years of his work,

6-1 Johannes Kepler.

Kepler tried to fit various combinations of circular motions, including epicycles, to Tycho's measured positions of Mars. He was unsuccessful. Of course, they could be fitted if he piled on enough epicycles.

But if he had as many epicycles as Ptolemy's system, what advantage was there to the Copernican theory? After Tycho's death,

ONE SEVENTH OF A DEGREE / 61

and at one point during these years of tedious work, Kepler developed a variation of the Copernican system which had quite a small number of epicycles. However, it disagreed with Tycho's by as much as one seventh of a degree—a very slight difference, to be sure. But Kepler did not believe Tycho's observations were wrong by even this small amount. So, he threw out this solution to his problem.

He gave up combinations of circles, probably remembering Tycho's comet orbits which were not perfectly circular. He simply decided to plot the positions of Mars and let them fall on whatever curve they might.

Look at Figure 6-2. Selecting one of Tycho's observations of Mars, Kepler drew the angle Tycho had measured between the lines of sight to the sun and to Mars (angle x in the left-hand drawing). Mars that night was out in space somewhere along the arrow-tipped line. The period of Mars is 687 days (we called it about 2 years in Chapter 4). Therefore, 687 days later Mars was back again at the same place in its orbit. The earth had meanwhile completed one revolution around the sun (365 days) and was 322 days along its second revolution (687 − 365 = 322).

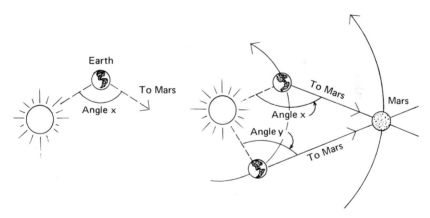

6-2 Determining a point on the orbit of Mars by observations 687 days apart.

Kepler's notes showed that the angle between the sun and Mars in the sky that day was y. He drew another line. He knew that Mars was somewhere out in space along that line. Somewhere? No, he knew just where: at the point where the two lines crossed. Mars was at the same place on its orbit on both nights. By using successive pairs of Tycho's observations, each 687 days apart, he drew successive positions of Mars. Hundreds of pairs of observations gave him a closely spaced series of points which outlined the orbit of Mars.

To his surprise, he found that it was almost a circle but not quite. The curve the points fit best is an *ellipse*. The ellipse, of course, was considered a rather unimportant curve, as compared to the perfection of a circle. In fact, Kepler had briefly considered ellipses as possible planet orbits some years before, but had discarded the idea because he shared the general prejudice that celestial bodies must move in circles—perfect circular curves.

The Imperfect Curve

The best way to define an ellipse is to describe how to draw one. Attach the ends of a short piece of string to two tacks, and put the tacks into a piece of paper on a drawing board, placing them fairly close together. Then push a pencil against the string, pulling the string tight, and slide the pencil against the string, all the way around the tacks, back to where you started. The closed curve you draw is an ellipse. The longer the piece of string, the larger the ellipse. With the same piece of string, the farther apart you place the tacks, the less like a circle it is. The closer the tacks are placed, the more circular the ellipse, until, if the two tacks were in

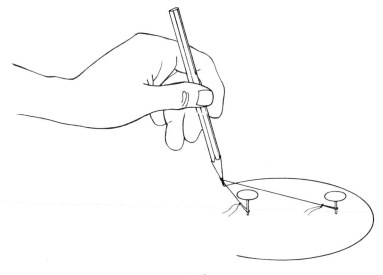

the same spot, you'd draw a circle. How much an ellipse departs from being a true circle is called the *eccentricity* of the ellipse. The larger the eccentricity, the more oval the ellipse. The eccentricity then of all circles is zero. With a 10-inch string, tacks 1 inch apart give an ellipse of eccentricity 0.1 (1/10), about the eccentricity of Mars' orbit.

The location of each tack you used to make the ellipse is called a *focus*. When plotting Mars in its orbit, Kepler found that the sun is at one focus. The other focus was empty. How surprising this must have been!

He also found that Mars' orbit did not form a perfect ellipse. Of course this could be partly due to inaccurate observations or measurements. But Kepler recognized that there is another way in which larger errors could have been made. If Mars does not move in a circular orbit, he reasoned, it is possible that the earth doesn't either. So Kepler tried ellipses of varying eccentricity for the earth's orbit and found that the observed positions of Mars fit well on his original ellipse if the earth's orbit was an ellipse, with an eccentricity of almost 0.02 (2/100).

If the earth's orbit is an ellipse, Kepler reasoned, then the *1*, called the astronomical unit (AU), that astronomers had taken for the distance of the earth from the sun must be a sort of average distance. In one part of its orbit, the earth is slightly closer to the sun; in the rest it is farther. Copernicus had thought of Mars' orbit as a circle with Mars always 1.5 AU from the sun, but Kepler saw that this figure was Mars' *average* distance from the sun.

In time, Kepler worked out the orbits for the other planets. All Tycho's observations fit extremely well if the planet orbits were drawn as ellipses with the sun at one focus. The average distance of each planet from the sun is about equal to the distance that Copernicus found for his circular orbits (column 1 of Table 2). The eccentricities are small, and different for each planet. No wonder astronomers had thought the orbits to be circular. You can see that all the planets' orbits, if drawn to a scale that would fit on the page, would appear to be circular, but with the sun off center in each orbit.

As you can see by the way an ellipse is drawn, the tack at one focus is off center by the amount of eccentricity of an orbit. So the sun is off center by a different amount for each planet's orbit, as shown in Table 2. These offsets are all in different directions, and are quite large for the outer planets. Six eccentricities (one for each planet) eliminated the forty-eight epicycles needed by Copernicus. And those epicycles were never needed again.

Table 2

	Approximate distances from sun (AU)	×	Approximate eccentricity	=	Distance of sun from center of orbit (AU)
Mercury	0.4	×	0.2	=	0.08
Venus	0.7	×	0.006	=	0.004
Earth	1.0	×	0.02	=	0.02
Mars	1.5	×	0.1	=	0.15
Jupiter	5.2	×	0.05	=	0.25
Saturn	10.0	×	0.05	=	0.5

The Laws of Planet Motion

Kepler replaced the idea of constant speeds in perfect circles with a different law of planetary motion. Figure 6–4 shows the orbit of Mars. B, C, and D are located 1/4, 1/2, and 3/4 of the distance around the orbit. If Mars moved at uniform speed, the

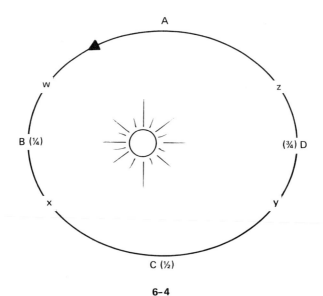

6–4

planet would go from *A* to *B* in about 172 days (687 ÷ 4 = 172). However, Kepler found that it goes beyond *B* in this time. On the other hand, Mars takes more than 172 days to go from *B* to *C* and also to travel from *C* to *D*. Then the planet goes from *D* to *A* in less than 172 days.

More than this, Kepler was able to show that every planet moves fastest in part of its orbit. He found that between *w* and *x* in Figure 6-4, when a planet is closest to the sun, it moves fastest. And it moves slowest between *y* and *z*, when it is farthest. He could pinpoint the speed change even more closely than this. Kepler expressed these facts in his *law of equal areas*: the line between a planet and the sun sweeps over equal areas in equal lengths of time, as shown in Figure 6-5.

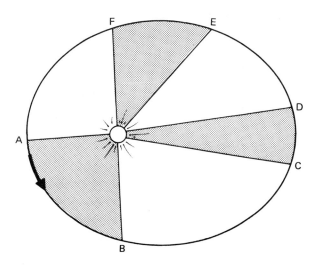

6-5 A planet moves most rapidly when it is at *A*, nearest the focus of the ellipse where the sun is. Its orbital speed varies in such a way that the planet moves through the shaded areas from *A* to *B* in the same amount of time it takes to move from *C* to *D* and *E* to *F*.

While working out the shape of the planets' orbits and their speeds, Kepler made another important discovery which is called *Kepler's Harmonic Law*. It is shown in Table 3. First, multiply the time it takes a planet to travel around its orbit (its period) by itself—P^2. For Mars, we multiply 1.8808 years × 1.8808 = 3.5374. Now multiply the distance of Mars from the sun (in AU's) by itself three times—r^3: 1.523688 × 1.523688 × 1.523688 = 3.5374.

They are both the same! For all planets $P^2 = r^3$ – a rather surprising thing. In a way, though, it is a more exact way of saying what Copernicus had found: the closer a planet's orbit is to the sun, the faster it moves along that orbit.

Table 3. Kepler's Harmonic Law ($P^2 = r^3$).

	Distance from the sun (AU) r	Period (years) P	Distance cubed r^3 =	Period squared P^2
Mercury	0.387	0.241	0.058	0.058
Venus	0.723	0.615	0.378	0.378
Earth	1.00	1.00	1.00	1.00
Mars	1.524	1.881	3.54	3.54
Jupiter	5.20	11.86	141	141
Saturn	9.54	29.46	868	868

The Copernican system as revised by Kepler is an extremely neat and simple theory. More than that, the sun definitely appears to be something special, while the earth is just another planet. After Kepler, the differences between Ptolemy's, Tycho's, and Copernicus' theories were no longer merely those of geometry or religious feelings, although very few people realized it at the time. As we have seen, all these systems could explain the motions of the sun and planets in the sky. Copernicus' was simplest to use, Ptolemy's offended people least, and Tycho's combined some of the advantages of both. But Kepler's work showed that there was a more basic difference between Copernicus' theory and Ptolemy's. His studies showed that motions of the planets were in some way dependent on the sun. Kepler's theory strengthened the argument for a sun-centered universe.

Of course, one could defend Tycho's theory and say that the planets move around the sun in ellipses while the sun, in turn, moves about the central earth in an ellipse. But then you have to explain the strange coincidence that the earth has exactly the same effect on the sun as the sun has on the five planets, in terms of varying speed along the orbit. And you would have to explain the further coincidence that the Harmonic Law, $P^2 = r^3$, applies to all five planets going around the sun and also applies to the sun going around the earth.

Today we would consider that the sizes of the sun, earth, and planets might have something to do with which one dominates the others. It was known in 1600 that the sun is larger than the moon and farther from the earth, but all estimates of its size were far too low. But in any case, few would have found it strange to think of a large body going around a smaller one.

Kepler's work accomplished two major steps that hadn't been taken in astronomy. The first was his acceptance of seemingly reliable data, even if the data did not agree with the theory he was trying to confirm. The usual practice had been to discard such data as incorrect or unimportant.

Perhaps more important are the kinds of questions Kepler asked. Other astronomers had been content to merely *describe* celestial motions. For most, the little wheels turning on big wheels was only a mathematical model. Therefore, they didn't concern themselves with what made the wheels turn. But Kepler asked why a planet's speed should vary in different parts of its orbit and why the outer planets' speeds were less than those of the inner planets. He even tried to explain why the planets followed ellipses. He suggested, among other things, there must be some attracting influence in the sun that pulled the planets around. In thinking about the physical cause of a planet's motion, he united astronomy and physics.

Kepler had to abandon the perfection of circular orbits and constant speeds. But in doing so he found a new kind of perfection and order in his laws of equal areas and his harmonic law.

Test Yourself

1. What is meant by the *eccentricity* of an ellipse?
2. What is expressed in Kepler's *Law of Equal Areas*?
3. What is expressed in Kepler's *Harmonic Law*?
4. Once Kepler found that the orbit of Mars is an ellipse rather than a circle, what other surprising thing did he find about the center of the ellipse?

5. What was the value of Tycho's observations to Kepler, even though Kepler did not accept Tycho's theory?
6. Why had other astronomers failed to notice that the orbits of planets were ellipses?
7. Did Kepler's discoveries show the reasons why the planets have the sun at one focus?
8. If any planet's orbit were a true circle, could Kepler's *Law of Equal Areas* still apply? Discuss.
9. Which orbit is less circular (and more elliptical), Mars' or the earth's?
10. How would you explain the fact that there are 3 more days between March 21 and September 22 than between September 22 and March 21?

7

Galileo's Telescope

In 1604, a Dutch spectacle maker put an eyeglass lens in each end of a tube. When he looked through the tube he saw distant objects appear nearer and larger, showing details invisible to the unassisted eye. The telescope, as this device was called, was an interesting novelty. Many people saw that it could be useful in warfare and navigation.

In the early summer of 1609, the news of the invention reached the ears of an Italian scientist, Galileo Galilei. Galileo immediately saw its importance to astronomy. He quickly bought two lenses from a local eyeglass maker and fitted them into a lead tube about 20 inches long. This telescope magnified objects about three times. It was about as effective as a modern pair of inexpensive field glasses and is called a refractor telescope. Amazed and delighted with what he saw as he turned it to the night sky, Galileo spent the next few months making a stronger instrument. By hand, he ground chunks of clear glass into lenses of the size and shape he wanted, and ended up with a telescope that magnified about thirty-two times. Using this instrument, Galileo made several important discoveries.

Galileo's Discovery

Galileo was the first person to use a telescope to view the heavens. At least, he was the first person to leave a record of what he saw, to carry on a series of observations, and to realize the important evidence the telescope revealed about the universe. If, indeed, he was actually the first person to turn a telescope to the night sky, he had an experience almost unique in the history of mankind. It may be compared to that of the first astronaut, al-

7-1 Galileo Galilei.

most 350 years later. Standing on the earth, Galileo had a new view of outer space. Traveling in outer space, the astronauts had a new view of the earth.

The moon was the first object Galileo looked at with his instrument. Until that moment, most astronomers believed that the moon was a smooth globe, clear and perfect as a celestial object should be. The dark markings—called the man in the moon—which everyone can see, had been explained in many imaginative ways. One was that in these patches the moon's material was so rare that it let the light pass through, rather than reflecting it back. Galileo had realized that these markings must be on the moon's surface. But his telescope revealed more than just a mottled surface.

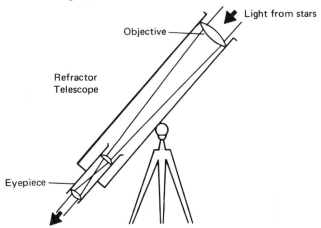

7-2 The objective lens focuses the light of each star near the bottom of the tube, where all the light rays cross at one point. Another lens (or two lenses, as shown here) form an eyepiece that magnifies the picture (images of several stars, or surface of the moon) when you look in the lower end.

He saw that the moon has great mountains and craters! Its surface is rough (see Figure 7-3) with large dark areas that he thought at first were seas. He saw that the moon resembles the earth; it is the same sort of object. Thus, the earth could be one of the celestial bodies; the Copernican system became more believable.

7-3 This photo was taken from an orbiting Apollo Command Module. In the foreground we see a closeup view of a portion of a rock-strewn, relatively fresh crater; in the background, a mountain range.

72 / ASTRONOMY

Next Galileo turned his telescope to the fixed stars. He found that it did not make them larger; they were still points of light. However, it did show many more stars—stars too faint to be seen by the eye alone. The telescope's lens brought more light to the eye to make them visible. Compare the number of stars in Figure 7-4 (left), the naked-eye view of the constellation Orion, with the number of stars in Figure 7-4 (right), a photograph of the same constellation through a telescope.

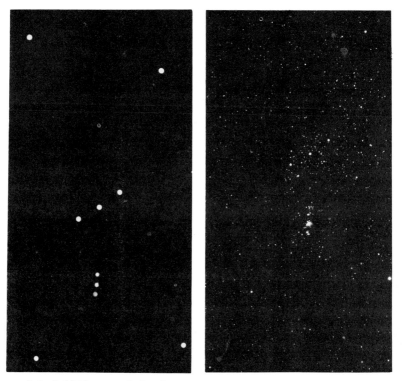

7-4 (left) The constellation Orion, prominent in the evening sky in winter. (right) The constellation Orion, showing stars not visible to the unaided eye.

Galileo's telescope also showed that some misty blurs of light in the sky are made up of many stars. One of these hazy patches, a star cluster called Praesepe, is in the zodiac constellation Cancer. You can find it in the sky with the help of Figure 2-7, page 19. Although it is not shown on the star map, it would be near Canis Minor in the diagram. A good pair of field glasses will give you a view of it like Galileo had with his telescope.

Surely, while stargazing, you have noticed the Milky Way. It is a wide band of faint light extending diagonally across the sky, like a stream of skim milk. Part of it is shown in Figure 2-7, page 19. It forms a background to the constellations Cassiopeia, Cepheus, and the Summer Triangle, and extends all the way around the celestial sphere. Galileo's telescope showed that the Milky Way is indeed, a multitude of faint, individual stars.

7-5 Telescopic photograph of the Milky Way in the area of the constellation Cygnus.

All these stars never seen before showed the unreliability of ancient writers, who didn't even know that these stars existed, yet whose opinions about the universe were generally regarded as the final word.

A Little Solar System

When Galileo turned his telescope toward Jupiter, he saw that the planet was round, like the sun and moon, and not a point of light, like the fixed stars. He noticed three very small but bright points of light near Jupiter, two to the east of the planet and one to the west. Oddly, these stars formed a straight line with Jupiter. The following night he was surprised to see that all three of the stars were west of Jupiter. The next night he saw only two of the stars.

Galileo looked at these new points of light through his telescope for many nights, sketching their positions in his notebook. A portion of his record is shown in Figure 7-6. You will notice that sometimes he saw four stars, sometimes three, and sometimes only two. He noticed that they were always near Jupiter, as that planet moved against the background of the fixed stars. These observations showed Galileo that the four stars revolve in orbits around the planet Jupiter. They are Jupiter's moons. As we watch each moon circle around Jupiter, it is sometimes behind the planet and sometimes at one side or the other. Sometimes it is in front of the planet, invisible against the bright disk. Galileo's observations showed that Jupiter's moons move with great regularity. He determined that the period of each one is different, varying from just under 2 to about 17 days.

7-6 Galileo's drawings of the moons of Jupiter.

Jupiter, with its moons, is a small-scale model of the Copernican system. It looks like the solar system would if we could view it from outside. And there is nary an epicycle in evidence! Galileo's discovery answered a common objection to the Copernican theory: that our moon is a strange exception because it alone moves around a planet instead of around the sun. It also proved, even more clearly than Tycho's comets, the error of the ancient doctrine that the earth is the only center of motion in the universe. It laid to rest the argument that if the earth were moving, the moon would be left behind because it could hardly be expected to keep up with the earth, rapidly orbiting the sun. Jupiter's satellites were having no difficulty staying with Jupiter!

Galileo was the first to see that Venus has phases, like our moon. As he observed night after night, he saw that Venus went through phases—shown in Figure 7-7 on page 76. This showed that Venus does not shine by its own light, but by reflecting sunlight. He believed that this must also, then, be true of other planets. On hearing this news, Kepler wrote Galileo that it now looked as if the fixed stars were suns, and the planets were, as he expressed it, earths.

With a good pair of field glasses or a small telescope, you can see what Galileo saw. But you must never look through field glasses or a telescope at the next object that Galileo examined. He looked at the sun with his telescope—a very dangerous thing to do, often causing permanent damage to the eyes. (In fact, Galileo went blind later on.)

What he saw is shown in the modern photograph in Figure 7-8 on page 77: the sun has blemishes. These *sunspots* are now known to be large, cooler areas on the sun that appear dark in contrast to the brighter and hotter solar surface around them. Sunspots are temporary, usually lasting only a few weeks to a few months. Large sunspots had been observed before Galileo's time with the eye alone, but were generally regarded either as something in the earth's atmosphere or something nearer the sun, silhouetted against its bright surface. However, Galileo watched the sunspots for many months and saw that they moved in a regular way across the disk to its edge, and then disappeared. He concluded from this that the sun must be rotating, carrying the current sunspots into our view and then out of sight again.

Sunspots, of course, were a striking exception to Aristotle's belief in the perfection of celestial objects. Moreover, the rotation of the sun made earth's rotation seem more possible.

76 / ASTRONOMY

7-7 These are four photographs of Venus taken over a 17-year period. Can you explain why Venus seems to get larger as it reaches the crescent phase?

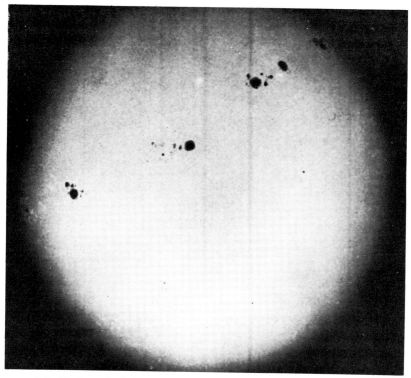

7-8 Photograph of the sun, taken on October 13, 1926 by W. W. Morgan with a telescope having a 12-inch lens.

Resistance to Galileo's Discoveries

Each of Galileo's discoveries was a blow to the Ptolemaic system and the ideas of Aristotle. But if Aristotle and Ptolemy had been standing at his elbow as he made these discoveries, they would probably have been eager to look through his telescope. Unlike their seventeenth-century followers, many of whom refused to do so, Aristotle and Ptolemy would probably have been the first to admit that they were wrong.

The system of the universe that Ptolemy and the earlier Greek philosophers had devised was based on their observations and explained these observations adequately. The scholars who taught the Ptolemic system as absolute truth in 1600, however, accepted it blindly. They looked on Galileo as a troublemaker, out to upset the established order and to question their authority. We look on him today as one of the founders of modern science, as one who reintroduced experiment and observation as the basis for scientific inquiry, rather than hashing over the opinions of authorities, with never a look at nature. Kepler, as we have seen, also considered measurements to be the final authority.

In 1610, Galileo published a book, *The Starry Messenger.* In it, he described his observations with the telescope. The book was widely read, but was most unpopular with professors and Church authorities. Primarily because of Galileo's book, high churchmen declared the Copernican doctrine "false and absurd." Anyone who taught that it was true was subject to punishment. Nevertheless, Galileo followed his first book with a second, *Dialogues Concerning the Two Principal Systems of the World.* This book was a crushing attack on Ptolemaic astronomy and a convincing argument for the Copernican system. He was so sure of his arguments that he wrote a long letter to one of the Cardinals in Rome saying that any other opinion was stupid. The eventual result was that Galileo was brought before the Inquisition in 1632 and threatened with torture, although he was 70 years old. This was about 25 years after publication of the *Starry Messenger.* He did formally "abjure, curse, and detest" his "errors and heresies." But as he left the court, he is reported to have said (in Italian): ". . . but the earth does move."

Part of the reason for his recantation was that his work was not yet finished. He still believed in his theories and spent the 8 remaining years of his life technically as a prisoner in his home, secretly writing his most admired and perhaps most valuable work, *Two New Sciences,* about motion and the structure of matter. He felt, rightly, that the truth or falsity of the Copernican system was no business of these authorities, who had never looked through a telescope, never observed a planet's motion across the sky, never drawn a planet's orbit, and never considered the evidence with open minds. He made his point. Scientific truth is decided by observation or experiment, and by reasoning sensibly about what you see. Two hundred years later, the early geologists, in a milder conflict with the authorities of established religion, put it this way: "We shall let earth herself tell us her story." And, armed with rock hammers, magnifying glasses, and a theory, they went to work to decipher the history and age of our planet, the earth.

Test Yourself

1. List five things that Galileo discovered with his telescope.
2. What is the Milky Way—in appearance and reality?
3. Galileo's telescope was of a certain kind still used today, and which has a certain class or type name. What is this name.
4. What is a *sunspot*?
5. In Galileo's (or any) telescope, stars remain points of light, but the nearer planets show as pale discs. Can you think of a possible reason why?
6. The moon's phases are easily explained by the moon's motion around the earth. But Venus also shows phases (as Galileo discovered) and yet it does not revolve around the earth. Can you think of an explanation?
7. In what main way did Galileo's discovery of mountains and craters on the moon weaken Aristotle's theory concerning heavenly bodies?
8. How did Galileo's discovery of Jupiter's moons help to confirm the Copernican view of the universe?
9. Why do you suppose that "full Venus" is smaller in the sky than "quarter Venus?"

Newton's Explanation

Galileo's observations and Kepler's calculations did make it seem likely that the Copernican theory was the true picture of the solar system. But a serious question remained unanswered. What had kept the earth moving around the sun during the long time that men had been observing the sky? When it was thought that the sun and planets and the sphere of the fixed stars were the ones in motion, there was no problem. It was generally believed that because they were celestial objects, they could move forever without effort.

Galileo's Experiment

The Greek philosophers had also taught that on earth rest is the natural state. According to them it took an effort—a *force*—to put something in motion and keep it moving. Galileo, rather than merely accepting this opinion, experimented with moving objects like the one in Figure 8-1. He found that when a wooden block is slid across the floor, it soon comes to rest. Friction with the floor slows it up. *Friction* is a resistance to the motion of one object sliding or rolling over a second object. Friction is the retarding force that slows down the wooden block in the experiment shown in the figure. When Galileo reduced the friction by using a polished block on a very smooth floor, the block slid farther before coming to rest. On smooth ice, this polished block slides still farther; there is even less friction. Galileo's experiments showed that the smaller the retarding force is, the less the block tends to slow down—and the farther it moves before coming to rest.

Although he couldn't prove it, Galileo reasoned that if all friction could be removed—if no resisting forces acted on the block—

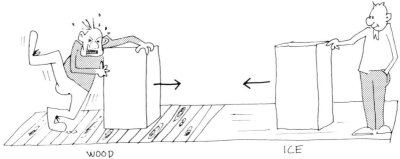

8-1 How far the block will go depends on how much friction there is. Which person's block will go farther if they both push equally hard?

it would continue in motion at the same speed forever. His experiments showed him that the Greeks were only half right: a force is required to start a motion. But not to keep something moving. In fact, Galileo found that force is also necessary to slow an object down, as well as to speed it up or change the direction in which it is moving. But a force is not necessary to keep an object moving in a straight line at a constant speed. Galileo's experiments convinced him that rest is no more natural than is straight-line motion at uniform speed. Galileo called this observation the *principle of inertia.* This principle states that an object at rest tends to stay at rest. Also, an object in motion tends to stay in motion—at a constant speed and in the same direction. A force is required to change the state of an object either at rest or in motion.

Galileo believed that there is no fundamental difference between celestial objects and earthly ones. (Remember in the last chapter that he discovered the moon is not a smooth and perfect sphere.) Therefore, his principle of inertia could apply to the movements of the planets as well as to blocks sliding along a smooth floor. Once started in motion, the earth and the other planets should keep moving. This principle would explain their continued motion for thousands of years. However, as you are probably already saying, it certainly doesn't explain why they move in ellipses at changing speeds, as Kepler found. Why don't they move as the principle of inertia states—at constant speeds in straight lines? Why do they keep going around the sun instead of moving along straight paths, which would long ago have taken them far away? If Galileo's principle is right, and if it applies to the planets, then

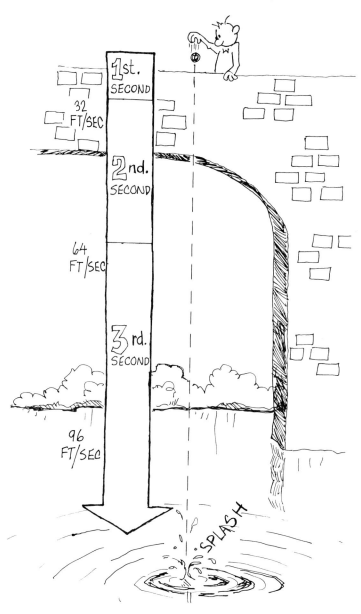

8-2 The downward speed of a falling object is increased by 32 feet per second each second that it is falling. The drawing shows the distances that a stone or piece of wood falls in successive seconds. Although it only falls 16 feet in the first second, it hits the water, 144 feet below the bridge, by the end of the third second.

there must be a force continually acting on them. What is this force? Galileo did not deal with this question, perhaps because he did not live long enough to do so.

However, Galileo did answer an old objection to the Copernican theory. People had long said that if the earth were moving, objects on its surface would be left behind. Birds flying through the air, and the air itself, couldn't keep up with the moving earth. Galileo pointed out that if you drop a stone from the mast of a ship lying at anchor, the stone falls to the deck at the base of the mast. If you drop the stone from the mast while the ship is moving, it still falls to the deck at the base of the mast. This shows that the stone shares the forward motion of the ship. If the stone didn't share this motion, it would land in the sea behind the moving ship, or at least on the deck behind the mast. Similarly, objects on the earth, birds in the air, and the air itself share the earth's motion. They are not swept off or left behind as the earth moves, any more than the stone is left behind the ship. A force has to act on the stone—you have to throw it, rather than merely drop it—before it falls anywhere but at the base of the mast. This is because it takes a force to change the forward motion of the stone, as the principle of inertia predicts.

The stone fell until it landed on something that could support it—in this case, the deck. If you stand in a field and throw a stone straight up, it always falls back to earth. Moving upward, it gradually slows down, stops, and then starts speeding up, back toward the earth. Galileo watched things fall. He found that freely falling bodies are *accelerated.* That is, they gain speed as they fall, as Figure 8-2 shows. Moreover, he observed that all of them speed up by the same amount in equal intervals of time. Their acceleration is always the same.

In addition, Galileo found that if, from the same height (see Figure 8-3), he dropped one stone and threw another one sideways, both hit the ground at the same instant. The harder he threw the latter stone, the greater was its speed as it left his hand, and the farther it went before it fell to earth in the same fraction of a second.

Apples and Moons

Galileo's data on falling bodies and his principle of inertia, as well as Kepler's elliptical orbits, were all known to Sir Isaac New-

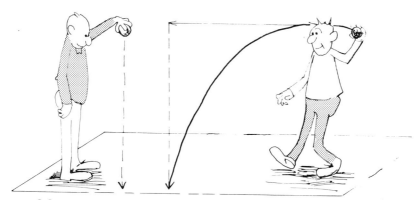

8-3 Both people let go of their balls at the same moment. The person on the left dropped his while the one on the right threw his sideways. However, they both hit the ground at the same time. Try this experiment yourself.

ton. Newton was born in England in 1643, the year after Galileo died. He graduated from Cambridge University at the age of 22 and went back to his home, where he spent the next 2 years trying to find out why the planets move eternally in elliptical orbits around the sun. An anecdote that almost everyone knows tells how he started on this problem as he watched an apple fall. Unlike the tale of George Washington and the cherry tree, this appears to be a true story.

Newton asked himself why the apple fell downward. Why didn't it go sideways, or upward? The earth, he thought, must attract it. Galileo had shown that the apple, or any other falling body, is accelerated—its downward speed steadily increases. The principle of inertia says that if there is no force on such a body, it moves at constant speed. Newton saw, therefore, that a force was acting on the apple. Because the apple fell toward the earth, it seemed likely that this force had something to do with the earth. It was an odd kind of force. Unlike a push on the apple given by the hand when you throw it, this "earth force" did not need to touch the apple. No hand or baseball bat or rope is needed to force the apple downward as it falls. Astronomers even before Newton had called this force of attraction *gravity*. But it was Newton who decided that it was the material of the earth pulling on the material of the apple. Since both the earth and the apple are made of matter, it seemed reasonable that the size of the force must depend on the *mass*—the quantity of matter—in both of them.

8-4 Sir Isaac Newton.

There would be a stronger pull on a larger apple and a weaker pull on a small crabapple. But Galileo had shown that large, heavy apples fall no faster than small, light ones. So Newton reasoned that a larger force is needed to accelerate a larger mass. That is, a 1-pound pull by the earth speeds up a large apple just the same as a ½-pound pull speeds up a smaller apple with half as much mass.

He also noticed that when he pulled on something like a rope tied to a wagon, the wagon pulled back on him. So he reasoned that the apple pulls the earth upward as the earth pulls the apple downward. Of course, the earth is much more massive than the apple. So the apple's ½-pound upward force scarcely moves the earth, while the earth's ½-pound downward force on the apple causes it to speed up 32 feet per second each second in its downward fall toward the earth.

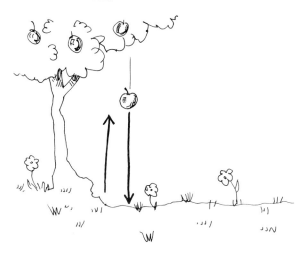

8-5 The earth is pulling the apple down while the apple pulls the earth up. The two pulls (forces) are equal and opposite. Which do you think has the greater acceleration? Draw a second set of arrows to show this.

Newton, like Galileo, believed that the same natural laws that apply to earth and apples also applies to the moon, the other planets, the sun, comets, and the stars. The movement of the planets in elliptical orbits showed him that a force is acting on them. Because they are material objects, a gravitational attraction can be expected between the sun and each planet. Yet the planets don't seem to be falling into the sun, like the apple falls to the earth. Newton recalled the motion of Galileo's stone thrown sideways as in Figure 8-3. As the stone moves forward, it is constantly falling toward the earth. The harder the stone is thrown, the farther it travels before reaching the ground. Newton could explain this curved motion in terms of gravity continually pulling the stone downward as it moved forward with the initial speed given to it by the thrower.

If the stone could be given a large enough forward speed, he reasoned, it would completely encircle the earth—always falling, but never reaching the earth's surface. Its curved path around the curved earth would keep them the same distance apart. If the moon had somehow been given a large sideways speed, then earth's gravity would keep it moving in an orbit around the earth. If the

88 / ASTRONOMY

planets were given fast enough motions, the sun's gravitational attraction would keep them moving in their orbits, always falling toward the sun.

Figure 8-6 shows the sun and a planet. From point *A*, the planet is moving in the direction of the longer arrow. At the same time, the force of gravity between the planet and the sun is pulling the planet toward the sun, as shown by the shorter arrow. The result is that the planet moves to point *B*. At *B* the planet tends to keep on moving in the new direction that took it to *B*. But the sun's gravity brings it down to *C*, and so on around the orbit. Points *A*, *B*, and *C* in this diagram should be drawn very close together, however, because at every instant, the force of gravity is accelerating the planet. Therefore, its direction of motion is constantly changing. At every instant, the gravitational attraction between the planet and the sun provides the proper force to accelerate the planet in its path.

Kepler had found that the planets orbit fastest when they are nearest the sun. This means that they are changing direction and speed most rapidly there—their acceleration is greatest. He saw

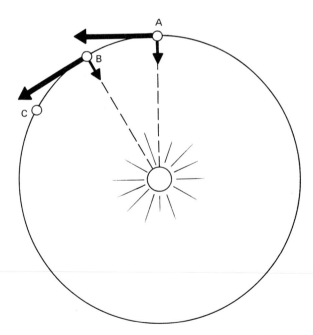

8-6 The force of the sun's gravity will keep a planet in orbit around the sun. Otherwise it would head out into space from point *A*.

that the planets in orbits close to the sun, like Mercury and Venus, are accelerated more than those farther from the sun. It was clear to Newton, therefore, that the force of gravity between the sun and a planet was somehow dependent on the distance of the planet from the sun.

The Effect of Distance

Newton saw his apple fall straight down along a vertical line to the center of the earth. As far as gravitation is concerned, a round object such as an apple, the earth, the moon, or the sun, acts as though all of its mass were concentrated at its center. Each member of the solar system, regardless of size, can be considered to have all its mass concentrated at a point. When an apple falls, the earth pulls on it as if all the earth's mass were concentrated at the center of the earth, some 4000 miles below the surface.

From his studies Newton concluded that every particle of matter in the universe attracts every other particle. The strength of the attraction depends on the amount of material in the objects (their masses) and the distance between them. The greater the mass, the greater the attraction. At double the distances, the attraction is four times less. These conclusions are known as the *law of universal gravitation.*

Newton said that this law applies to "every particle of matter in the universe." Of course, he couldn't prove it. The law says that the kitchen stove and the kitchen table attract each other, that two people passing on the sidewalk attract each other. Although they don't seem to, Newton was confident that this was because the gravitational forces between two such small objects are so very tiny. The smaller forces are swamped by other much larger forces, such as the earth's gravity. His confidence was justified. About 130 years later, careful laboratory measurements did reveal the tiny gravitational attraction between two small lead spheres.

Newton couldn't prove that the law of inertia and his law of gravitation applied to the whole universe—to the fixed stars and whatever might lie beyond them. But all the bodies in the solar system moved as these laws predicted, and he saw no reason to believe that other celestial objects were different. Newton's triumph, and the usefulness of his work to us today, are in his set of three laws. They explain and predict both the movement of planets and the moon and falling of apples or anything else.

Test Yourself

1. What does the *Principle of Inertia* state about the motion of objects?
2. What is *friction*?
3. Why did Galileo believe that inertia would keep the planets moving once they got started, even though every moving object he had tested eventually slowed to a stop?
4. What did Galileo learn about the speed of falling bodies as they fell toward the earth?
5. What two simple ideas make up Newton's *Law of Universal Gravitation*? His law of inertia?
6. Newton reasoned that if a thrown stone could be given enough forward speed, it would encircle the earth, always falling, but never reaching the earth's surface. Has anything happened within the past 20 years to confirm this idea?
7. Since Newton couldn't prove his universal law of gravitation (that is, prove that it applied everywhere in the universe) why should he have trusted it?
8. Has anything happened in your lifetime to support or confirm the idea of universal gravitation? If so, what else does it support?
9. How did Newton explain Kepler's law of areas?
10. How did he explain the fact that the inner planets travel along their orbits at a higher speed than the outer ones?
11. A car turns a corner at a constant speed of 30 miles per hour. Does this mean that no force is acting on it?
12. How did Newton explain Galileo's observation that all falling objects on earth fall at the same speed?
13. A thin, light object (like a feather or a parachute) falls more slowly because another force (air resistance) is able to oppose its fall. Does this make Galileo's principle of inertia wrong? Or does it support it?

"But the Earth Does Move!"

In about 150 years, five men had erased the picture of the universe that had been accepted almost without question for more than 14 centuries. Copernicus' idea, Tycho's accurate measurements, Kepler's persistence, the imagination and clear thinking of Galileo, and Newton's brilliant theory had replaced the time-honored picture with a new one. The idea that the earth revolves around the sun, considered ridiculous in the sixteenth century and immoral in the seventeenth century, came to be taken as a matter of course in the eighteenth century.

In the nineteenth century, parallax of the stars—sought for since Ptolemy's day as proof that the earth is moving—was at last observed. But it was almost an anticlimax. By then, astronomers had made many new discoveries about the universe, all of them consistent with Newton's laws. Some of these discoveries had even been predicted on the basis of these laws. And in 1727, an English astronomer named James Bradley presented proof of the earth's motion that no one had even thought of before.

Aberration

On a calm, rainy day, as you wait for a traffic light to change, raindrops fall straight down past the car window. But when the car starts moving again, the drops seem to be falling somewhat diagonally past the window, as if they were coming from a slightly forward direction. When you stop the car at the next light, however, the rain is falling vertically again. The raindrops were falling straight down all the while, but an effect called *aberration* makes them appear to slant toward the rear of a moving car.

91

Bradley may have noticed the same effect from a stagecoach window. At any rate, aberration of raindrops suggested to him a way of proving whether the earth is moving. An analogy, illustrated in Figure 9-1, shows his line of argument. If you stand still, holding a length of hollow pipe vertically, a raindrop which enters the pipe will fall straight through it (left). But if you start walking, the raindrop can't fall straight through the pipe. It hits the inside of the pipe before it reaches the bottom, as shown in the middle diagram. However, if you tilt the moving pipe just right, as in the right diagram, the raindrop falls all the way through the pipe. If you are walking in the opposite direction, the tilt must be reversed. If you are walking faster, it must be increased.

9-1 An example of aberration of raindrops.

A telescope catching light from a star can be compared to a pipe catching raindrops. If the earth isn't moving, then you can point the telescope directly toward any star and its light will move down the telescope tube and reach your eye. But if the earth is moving, you must tilt your telescope a little toward the direction in which the earth is moving, or the light won't travel all the way down the tube.

The earth's movement in an almost circular orbit, Bradley saw, would mean that the direction in which the telescope is tilted from a star's true direction would change continually throughout the year. And because there are stars in all directions, the tilt wouldn't be the same for all of them at any one time. Neither the direction nor the size of the angle of tilt would be the same for all stars.

Light from stars 90° north of the ecliptic, for instance, comes straight down on the earth's orbit, like the raindrops in Figure 9-1. For those stars (see Figure 9-2), the tilt away from the true direction would be largest. Stars dead ahead, toward which the earth is moving on a given night, would not be shifted at all. The light from other stars comes in at various angles to the direction of the earth's motion. The amount of tilt necessary for each one would be different, but predictable.

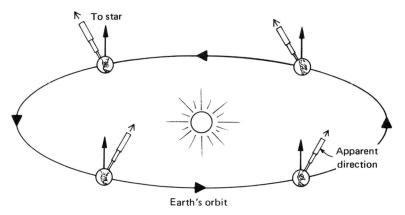

9-2 A person looking through a telescope at a star 90° north of the ecliptic will have to point his telescope in the directions indicated. (The amount is greatly exaggerated in the drawing.)

If the earth is moving in an orbit, all the stars will appear to be shifting position on the celestial sphere continually. Each star will appear to move by a slightly different amount in a slightly different direction. To see whether the stars did shift position this way, Bradley had to measure angles between many stars on many nights. He had to measure very precisely, for the changes were very small. In the end, Bradley had presented proof that stars do appear to move back and forth each year relative to one another, in the amounts and directions predicted for aberration. Thus, all the evidence fitted together. The orbital motion of the earth appeared to be a reality.

Stellar Parallax

As the Copernican idea came to be generally accepted, the fact that no star showed parallax began to be considered evidence for the tremendous distances of the stars, rather than meaning that the earth does not move in a yearly orbit. As we saw in Figure 5-3, page 53, the farther away an object is, the smaller is its parallax. Nevertheless, as stronger and stronger telescopes were made, the search for stellar parallax continued. In 1838 three astronomers, working independently, were able to measure the parallaxes of three different stars. Thomas Henderson, at the Cape of Good Hope in Africa, found the yearly angle of parallax of a star called Alpha Centauri to be extremely small, only 1/2400 of a degree. It is no wonder that Tycho couldn't find stellar parallax without a telescope—Alpha Centauri has just about the largest parallax of any star—it comes very close to being the star nearest the earth. (Because it is only 30° from the south celestial pole, it is not visible from the United States or from Europe.)

The other two men measured the yearly parallax of different stars. F. W. Bessel, in Germany, found the parallax of a star (visible only through a telescope) in the constellation Cygnus, to be even smaller. Wilhelm Struve, in Russia, measured that of Vega, which was smaller still.

Since then, parallax measurements have been made on about seven hundred stars. But these first measurements—or, indeed, any one of them—prove that the earth orbits the sun. The apparent movements of stars repeat themselves annually, and the greatest differences in position are between observations made 6 months apart. It takes the earth 6 months to move from one side to the opposite side of its orbit. The different amounts of parallax of these three stars show them to be at different distances from the earth. Vega, with a parallax six times smaller than that of Alpha Centauri, is six times farther away.

The fact that parallax differs so widely shows that the stars are scattered through space, rather than pasted on a celestial sphere, all at the same distance. As we said in Chapter 1, the sky appears to be a half sphere and the horizon, a circle. This is because we see equally well in all directions. The stars only *appear* to lie on a sphere. Their positions in the sky are not definite locations, but merely directions, or lines of sight, to each star. The flashing lights of an airplane can have the same position on the celestial

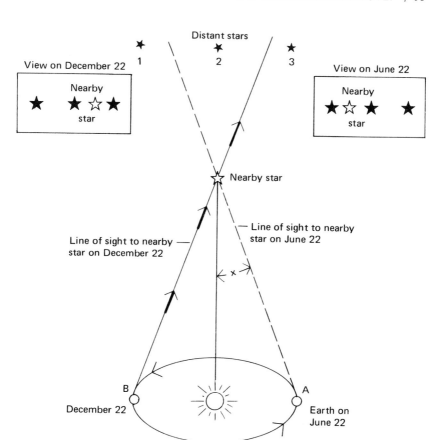

9-3 When you view the nearby star from position *A*, it will appear to be between stars *1* and *2*. However, 6 months later, when the earth is at *B*, the star seems to be between stars *2* and *3* in the sky. The difference in the lines of sight to the nearby star from these two positions on earth (angle *x*) is the stellar parallax. (In the diagram the angles between the two lines of sight are greatly exaggerated.)

sphere as a far-off planet or an even more distant star, if all three are in the same line of sight.

Although astronomers describe the direction of an object in the sky in terms of its location on the celestial sphere, after 1838 the sphere could no longer be considered a reality. And a sphere that isn't there can't rotate. The observed daily motions in the sky must be due to the earth's daily rotation. The measurements of stellar parallax certainly suggest that the earth is rotating, as well as revolving yearly about the sun.

The Foucault Pendulum

Galileo had also shown that the earth *could* be rotating. But the fact that things aren't flying off the earth's surface or being left behind doesn't really prove or disprove rotation. In fact, until 1851 there was no direct evidence that the earth rotates. In that year a French physicist named Jean Foucault (*fu*-koh) fastened one end of a long wire high in the dome of a building in Paris and attached a heavy metal ball to the other end. This ball and wire formed a long pendulum, somewhat like the much smaller pendulum of a grandfather clock. The base of the metal ball had a sharp point, and Foucault had placed a ring of loose sand on the surface beneath it. He started the pendulum swinging by drawing the ball toward him and letting it go. The pendulum began to swing back and forth. It crossed and recrossed the ring of sand, and at each crossing drew a straight groove across the sand.

Once he had set the pendulum in motion, the only force acting on it, except for a very small amount of air resistance, was gravity. And this force was, of course, in a downward direction—not sideward. Air resistance would slow the pendulum a little, but there was no force to change its direction of swing. So, according to the principle of inertia, the pendulum should have continued to swing *in the same direction.*

Yet it soon became apparent that the swing of Foucault's pendulum was shifting gradually to a new direction. He saw the line across the sand fan out, as shown in Figure 9-4, starting along line *Aa*, and gradually moving to *Bb*, then to *Cc*, and so on. The plane of swing had shifted with respect to the floor of the room, to landmarks outside, and to compass directions. Note that the end of the pendulum swing drifted to the right, or east, in Paris.

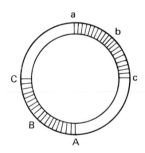

9-4 Foucault's pendulum marks in a ring in the sand.

"BUT THE EARTH DOES MOVE!" / 97

9-5 Rotation of the earth was proved by Foucault using a pendulum (top). The ball of the pendulum knocks down a peg every few minutes (bottom).

98 / ASTRONOMY

Foucault concluded that the pendulum only appears to change its direction of swing. It appears to do so because the earth is turning underneath it, carrying it along. (See Figure 9-5.) Imagine swinging a chain and locket (the pendulum) back and forth over a phonograph record (the earth) as it plays. The swing of the chain keeps the same direction with respect to the furniture in the room (the fixed stars). However, an ant on the record would see the line of swing changing direction continually with respect to marks on the label underfoot (a landmark on the earth). It is the record which is moving—not the room and not the line of swing of the locket.

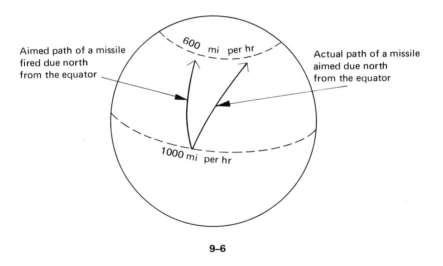

9-6

The Military Evidence

By the mid-nineteenth century, long-range artillery furnished another indication that the earth rotates. Cannon shells, fired in the Northern Hemisphere always veer off to the right, as shown in Figure 9-6. This fact must be considered when the cannon is aimed. This can be explained in terms of Newton's laws and a rotating earth. If a missile is fired due north from the equator, it starts its trip with the northward velocity given it when it was launched. But it also starts the trip with an eastward velocity of about 1,000 miles per hour—the velocity it shared with the rotating earth before launch. After launch, the only force acting on the missile is gravity, which pulls it directly downward. Neither the northward nor the eastward velocity is affected (except for the slight effect of air resistance). As the missile moves northward, it passes over areas which are closer to the earth's axis of rotation where the speed of rotation is less. The farther north the missile goes, the greater is the difference between its original eastward velocity of 1,000 miles per hour and that of the earth under it. So, relative to the ground, the missile veers off to the east, or to the right of the direction of aim.

Test Yourself

1. Give a simple definition of *aberration* as used in this chapter.
2. Vega, Alpha Centauri, and Bessel's star all have different parallaxes as measured from the orbiting earth. What does this show?
3. What do parallax measurements reveal about the celestial sphere? The earth?
4. Why are parallax differences (apparent changes in star positions) greatest between observations taken 6 months apart?
5. According to the principle of inertia, Foucault's pendulum should always swing back and forth in its original direction. Yet, over a period of hours it changes direction with respect to the walls of the room. Does this pendulum violate the inertia principle? What is really happening?
6. Why couldn't Tycho Brahe make measurements of stellar parallax?
7. If even today, with extremely strong telescopes, no star had yet been found to show parallax, would this prove that the earth is *not* moving around the sun?
8. Which of these proves that the earth is revolving around the sun: the Foucault pendulum, aberration, or stellar parallax?
9. If Foucault's pendulum were swinging in Australia or Argentina, or some other place on earth south of the equator, would it behave the same way?

10

Weighing the Earth, Sun, and Planets

Mass is the total amount of material in a body. It is not the same thing as weight. The weightless astronauts orbiting the earth still look the same; their masses haven't changed. Your weight is the gravitational force between the earth and you. It is, however, directly related to your mass. When you reduce your mass (by dieting, for instance), you reduce your weight also.

Centers of Mass

According to Newton, as the earth pulls the apple downward, the apple pulls the earth upward. These two forces are opposite but equal in amount. Because the earth is so much more massive than the apple, the earth's acceleration by the apple is too small to be measured. What if the two attracting objects are equal in size and mass? Then they will be accelerated equally toward each other. And they will collide at a point midway between their centers. This point is called the *center of mass*.

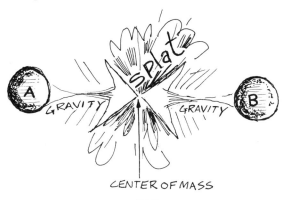

10-1

Now let us suppose that instead of merely falling, these two equal masses, *A* and *B*, have a certain sideways speed, as Figure 10-2 shows. *A* is continually falling toward *B*, and *B* is continually falling toward *A*. But both masses are moving sideways as well. The result is that they revolve around their center of mass, one opposite to the other. If they are started with the correct sideways motion, they will move in the same direction around the same circle, always on opposite sides of the center of mass.

But let's think about a much, much bigger mass than the apple, such as the moon. It pulls the earth much more than the apple. But since they do not crash into each other like the falling apple

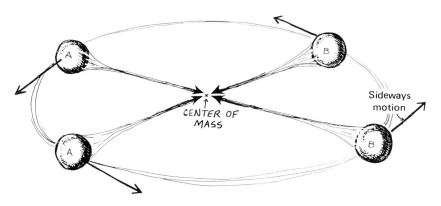

10-2 Objects of the same mass will orbit around their center of mass if given a sideways speed in addition to the force of gravity between them.

and the earth, both the earth and the moon must have sidewise motions. Look at Figure 10-3. When one of the two bodies is more massive than the other, the center of mass is not midway between them. Suppose *B* had twice the mass of *A*; then the center of mass is exactly twice as far from *A* as from *B*, and *A* would follow a circular orbit twice as large as B's circular orbit. The center of mass is always nearer the larger of two masses. If *B* is much larger than *A*, its orbit becomes much smaller, and *A* must go much faster than *B* in order to stay on the opposite side of the center of mass. If *B* were large enough, it is possible that the orbit of its center might be smaller than *B* itself. And if *A* were small enough then it would look as if *A* were orbiting around *B*, although in fact, both are circling their common center of mass. Only a slow wobbling motion of *B* would reveal that it is moving in an orbit.

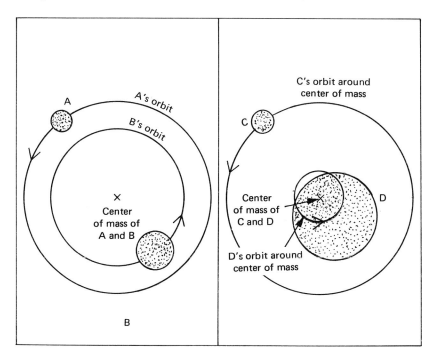

10-3 (left) *A* revolves in a large orbit about the common center of mass while *B* revolves in a smaller orbit around it. *A* must also move 1½ times as fast as *B*. (right) *D* is so large that the center of mass of *D* and *C* is inside *D*. Thus, it looks like *C* is revolving around *D* while *D* only wobbles. *C* must move four times as fast as *D*.

Do we find any motions in the solar system that show this? Yes, in our own back yard. We can see the monthly wobble of earth reflected in our view of nearby objects such as Mars, which doesn't appear to move among the stars in the sky at quite the speeds predicted. During the course of each month, as the moon swings the earth around the earth–moon center of mass, the earth gets a little behind schedule in its orbit around the sun. So when we look out at Mars, it appears to gain speed and get ahead of where it should be in the sky. Later in the month the earth swings ahead of schedule and Mars appears to slow down and get behind where it should be. This small lead and lag repeats itself each month.

This is how we know the earth is revolving in a small monthly orbit. In Figure 10-4, the moon (about 240,000 miles away) moves around the earth in a large orbit each month, while the center of the earth moves around in a tiny orbit. The center of both of these orbits is the center of mass of the earth–moon system. Because the earth is much more massive than the moon, the center of mass lies inside the earth. And it is this center of mass that travels around the sun in a yearly orbit. The earth and the moon, always on opposite sides of the common center of mass go around it twelve times during each year. From the earth, wobbling in its small monthly orbit, we see Mars a little ahead of its predicted position when the earth is on one side of this little orbit and a little behind when the earth is on the other side.

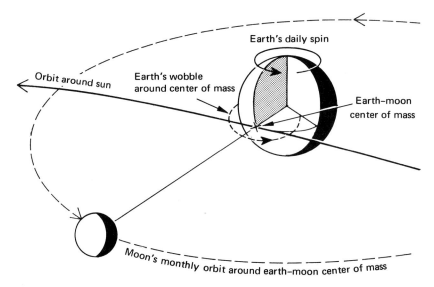

10-4 The center of mass of the earth–moon system is within the earth itself.

A motion predicted by Newton's laws has been observed—a wobble in the earth's orbit due to the moon's pull, first one way, then another. The earth doesn't go smoothly around the sun. The center of mass of the earth–moon system does. You have probably noticed that this wobble (and the moon's larger orbit) is like Ptolemy's epicycles—wheels going around on wheels. But notice that the modern "epicycles" are on orbits around the sun (not the earth) and that the earth's epicycle (wobble) affects all observations by astronomers, who must correct for this monthly wobble. The earth-moon wobble may seem to you to cause a lot of trouble, but it is worth studying for one important measurement—the mass of the moon, which is determined from the sizes of these monthly orbits of the earth and moon.

The center of the earth is 3,000 miles from the center of mass. The center of the moon is 240,000 miles from it. Thus, the center of the moon is 240,000 ÷ 3,000 = 80 times farther from the common center of mass than is the earth's center. Therefore, the mass of the moon is about 1/80 that of the earth.

Newton's laws also predict that the sun's center isn't the center of the orbit traveled by the earth-moon system each year. We would expect this system and the center of the sun to travel around their center of mass, the sun moving in a small orbit while the center of the earth-moon system travels in a larger one. Each planet and the sun should move in a pair of orbits. So the sun must be pulled every which way as the planets move around it. Thus, the sun should wobble too, but even the most precise instruments can only barely detect it. This is because the sun's mass is so large compared to that of the planets.

Masses in the Solar System

If you knew that six doughnuts cost twice as much as three, you wouldn't know how much six or three of them cost unless you knew the price per doughnut. Newton knew that the earth's mass pulled on the apple and accelerated it 32 feet per second, each second, at a distance of 4000 miles from earth's center of mass. He knew that it accelerated the moon 1/100 foot per second, each second, at a distance of 240,000 miles. But he did not know how much a one-pound or a one-ton mass at those distances—or any other—would accelerate the apple or the moon. This meant that he did not know the strength of earth's pull on the apple. This meant that he didn't know the mass of the earth.

Then in 1798 Henry Cavendish, an English scientist, was able to measure the tiny acceleration that a lead ball of known mass had on a much smaller ball an inch away. He compared the acceleration that the earth gave the small metal ball with the acceleration that the lead ball gave it. By taking into account the difference in distance between the small ball and the lead ball (1 inch) and between it and the earth's center (4000 miles), he was able to determine the mass of the earth. It turned out to be 600 billion billion tons. Now the mass of the moon (1/80 that of the earth) could be put into tons or pounds.

Back in the sixteenth century, the sun's diameter had been estimated to be several hundred times that of the earth. After Galileo's time, with larger telescopes, the parallax of Mars, whose orbit lies ½ AU outside the earth's orbit, could be measured. Mars' passing distance from the earth turned out to be 46½ million miles. Therefore, *1 AU*, the distance of the earth from the sun (the astronomical unit) is 93 million miles.

Once the sun's distance was known, its size could be determined. The sun's diameter in the sky is ½°. By the same methods as modern surveyors use, it can be calculated that the sun is 865,000 miles across—or about 108 times the diameter of the earth. This means that its volume is more than a million times that of earth. The sun could be made of material thousands of times lighter than that of the earth and still be much more massive than the earth.

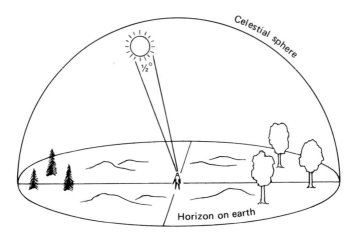

10-5 Astronomers measure the angular size of a large object in the sky, such as the sun, by extending the lines of sight from observer to the opposite sides of the object and measuring the angle between them.

After Cavendish's measurement, the mass of the sun pulling the earth, 93 million miles away, could be found out. Each second the sun accelerates the earth in its orbit only about 2/100 foot per second. But when you take into account the large distance, this shows that the sun is a whopping 2 billion-billion-billion tons, about 333,000 times the earth's mass.

These large numbers are confusing, so astronomers use the earth's mass as a unit and give the masses of the planets and moons in terms of the earth's mass (see Table 4, page 124).

10-6 Venus, above, and Jupiter with four bright moons photographed with a 10-inch telescope.

The mass of any planet that has moons (see Figure 10-6) can be measured in the same way. The orbit of a moon can be detected from its movement around the planet in the sky. Its period (how long it takes the moon to circle the planet) can also be determined. Then we know how fast it is moving and its acceleration toward the planet. This tells us how great the gravitational force of the planet is, and hence its mass. The mass of Jupiter, the largest planet, turns out to be 317 times the earth's mass. Calculations made in this way from any of its twelve moons give the same result.

The Planets Attract Each Other

From Newton's laws, we know that if every particle of matter in the universe attracts every other particle, then the planets must attract each other. The amount of gravitational force between any two planets depends, of course, on their masses and the distance between them. And because each planet is orbiting the sun at a different average speed, the distance between any two planets is constantly changing.

As you recall, when Kepler plotted the positions of Mars in its orbit, they didn't exactly fit an ellipse. When he replotted them, taking into account the fact that the earth's orbit was also an ellipse, they fitted quite well. However, many positions of Mars against the background of the stars still continued to be a bit off. Newton's laws explain this and, in fact, predict it. The gravitational attraction of the massive planet Jupiter pulls Mars off course slightly, or *perturbs* it. When the two planets are closest together (on the same side of the sun), the effect is greatest. When they are on opposite sides of the sun, this perturbation is least (Figure 10-7).

Jupiter is perturbed by the smaller Mars too, but not to the same extent. Astronomers can figure out the mass of one planet

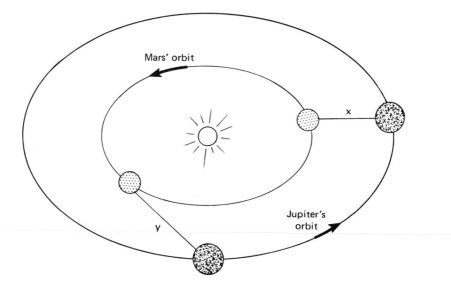

10-7 Jupiter will perturb Mars more when they are close, as at *x* than when they are farther apart as at *y*.

from its perturbing effect on another. Thus, the mass of Mars can be determined from the perturbing effect it has on the earth. And its mass, determined in this way, agrees with the mass determined by the acceleration of its moons. Venus' mass can be determined by its perturbing effects on the earth and on Mercury—the only ways to do it, since Venus has no known moons.

Of course, all the planets affect each other. And the results of the interaction of all these masses is a difficult mathematical problem. Modern electronic computers are a great aid in solving the problem of the mutual gravitational attraction of all the planets and in predicting planet orbits accurately enough for space probes to visit them. (If we didn't know Mars' mass, we couldn't calculate a space probe's approach to it, or how to land the space probe on Mars.)

In studying the universe, astronomers are naturally interested in the amounts of material—the masses—of all the things they can see. Newton's laws provide a method of doing this. Whenever one object near a larger one is seen to orbit around the larger one, like a planet orbits around the sun, the mass of the bigger object can be calculated from the acceleration of the smaller one. Or if one object's motion is perturbed by another, like Venus perturbs the earth in its orbit around the sun, then the mass of the perturbing object can be estimated. In later chapters we will use these methods to estimate masses of everything we see in the universe.

Test Yourself

1. What is *mass*?
2. What is *weight*?
3. How does mass affect weight?
4. Where is the center of mass of the earth–moon system?
5. The gravitational attraction between two planets depends on what three things?
6. The earth does not move smoothly around the sun in its yearly orbit. What does move smoothly?
7. Can you explain the evidence mentioned in this chapter that the earth, like the moon, revolves monthly about the center of mass of the earth–moon system?

8. What is meant by the statement that Venus *perturbs* the orbit of the earth?
9. Because all planets attract the sun gravitationally, it makes sense to expect the sun to wobble slightly over a period of time. Why is this wobble so difficult to detect? Why did we say "over a period of time" rather than in a yearly orbit?
10. Is any place in the universe beyond the effect of earth's gravity?
11. The mass of the moon is 1/80 that of the earth. Does this mean that an astronaut walking on the moon weighs 1/80 of what he does on earth?
12. Why doesn't Mars perturb Jupiter as much as Jupiter perturbs Mars?
13. In addition to the acceleration of falling bodies, what else is needed to measure the earth's mass in tons?
14. Why was a telescope needed to measure Mars' parallax while Tycho measured the moon's parallax before telescopes were invented?

11

Uranus, Neptune, and Pluto

In 1757, a 19-year-old German musician named Friedrich Wilhelm Herschel fled to England to escape military service in the Seven Years War. A decade later, after a great deal of hard work, he was comfortably settled in the resort town of Bath. Everyone knew him as William Herschel. He was a capable and popular music teacher and director of the city's orchestra. In 1772, a friend loaned him a telescope, and, as Herschel said, "this opened the kingdom of the skies" to him. Herschel became an enthusiastic amateur astronomer. He also became a telescope maker because he found good ready-made telescopes too expensive. In his spare time, he tried first to make a *refracting telescope*, which uses glass lenses to produce a magnified image. However, he couldn't make lenses good enough to satisfy him.

Then he began to make a *reflecting telescope*, using mirrors instead of transparent lenses. The first successful telescope of this type had been built by Sir Isaac Newton in 1668. In a reflecting telescope a curved mirror at the bottom of the tube takes the place of the lens at the top of the tube. This mirror reflects the light from a star back up the tube. An image of the star is formed near the front end of the tube, where the reflected rays come to a focus. This focus is in the way of the incoming light, of course. But Newton had solved the problem by placing a flat mirror, mounted diagonally, in the middle of the tube, as shown in Figure 11-1. This mirror reflects the light to the side just before it reaches the focus. Thus, the light comes to a focus outside the telescope tube, and can be viewed through a small lens (the eyepiece). Newton's curved mirror performs the same task as a lens (focusing light) but is easier to make and mount in large telescopes than is a transparent lens.

112 / ASTRONOMY

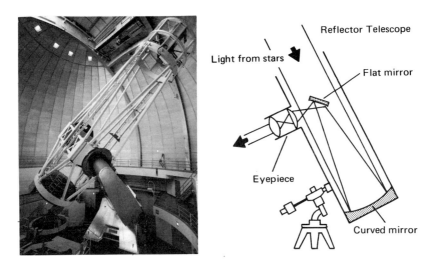

11-1. The big 120-inch reflecting telescope at Lick Observatory has an open framework rather than a tube as in the diagram above.

Until Herschel made them popular, few reflecting telescopes were used. Good telescope mirrors are difficult to make. Before Herschel made a satisfactory one, he discarded about two hundred. But, after a few years' work, he produced a really excellent instrument, 7 feet long, with a mirror 6½ inches in diameter. Using this

11-2 The first reflecting telescope, built in 1668 by Newton.

telescope, Herschel began to carefully map the sky, piece by piece, constellation by constellation, including every star he could see. He could measure and plot their positions in the sky relative to each other very precisely. Herschel was helped in this work, and indeed throughout his career, by his sister Caroline, the first woman astronomer. She used the telescope and did many tedious and difficult calculations.

11-3 Caroline Herschel.

By March 1781, Herschel had reached the zodiac constellation Gemini (The Twins) in his survey of the sky. One night he was studying some stars in Gemini when his eye was caught by a star brighter than the others. He was surprised to find that it was not shown on the older, less detailed star maps and star lists that he was using as a guide. Also, it appeared to be a disc, rather than a point of light. He thought that it was probably a comet.

Herschel's "Comet"

The eyepiece that Herschel was using magnified objects 227 times. He exchanged this eyepiece for a stronger one, increasing the magnification to 460 times. The diameter of his "comet" image doubled, while the stars still looked like points of light (as they should in a good telescope). Then he changed eyepieces again. Now the images were magnified 932 times. The star images were little changed, but the "comet" seemed twice again as large. Now he was certain that it was not a star.

He watched the "comet" for a month, and saw that it moved in a regular way relative to the stars. In early April he sent notices of his discovery to astronomers all over Europe, asking them to track it. Soon it was being observed nightly by many skilled astronomers using the best telescopes in the world. Among them was a German astronomer named Johann Bode at Berlin Observatory, whose map of the comet's movement in the sky is shown in Figure 11-4.

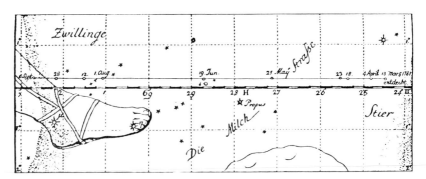

11-4 Johann Bode's diagram of the apparent path of Herschel's "comet" from observations between March 13 and September 13, 1781 (the line with the small circles and dates). The blocked line parallel to it and slightly south is the ecliptic (the apparent eastward path of the sun around the celestial sphere). "Die Swillinge" is the German word for "The Twins"; "Die Milch Strasse" is the Milky Way.

By the end of May the moon had moved so close to the "comet" on the celestial sphere that it was visible only for a few minutes after the sun went down. It couldn't be seen during most of June. Then, late in the month, it appeared in the sky just before sunrise. If it were a comet, traveling in a long, oval orbit around the sun, this is the very time that it should have developed a tail. From ancient times it had been noticed that all comets develop a tail as they near the sun. But it didn't. In addition, Herschel's "comet" looked different than other comets. It wasn't fuzzy around its edges; it was a clear, sharp disk like Jupiter or Venus. And although its eastward movement among the stars increased as it neared the sun, it didn't increase nearly as much as comets do. In fact, it seemed to move eastward more slowly than the sun. Furthermore, there was no change in its apparent diameter, as there would have been if it had been coming closer to earth in a comet-like orbit.

By November 1781, part of its orbit around the sun could be plotted. The full orbit would completely encircle that of Saturn, the farthest known planet. Herschel's new planet, as it now appeared to be, was double Saturn's distance from the sun. So it was not near the sun at all, but only in the same part of the sky. Kepler's harmonic law ($P^2 = r^3$; see Table 3, page 66) indicated that it would take 80 years to travel once around this orbit. Astronomers also figured out that this new planet's diameter was more than 3½ times that of the earth. The new planet was named *Uranus*. In Greek Mythology, Uranus was the father of Saturn, who was the father of Jupiter.

The discovery of Uranus wakened a popular interest in astronomy that was never equaled until after Sputnik I was launched in 1957. Social and official London rushed to do Herschel honor. His front yard was crowded with visitors. The way to his observatory was blocked by carriages filled with curious and admiring throngs. Words like *orbit, perturbation,* and *planet* were heard for the first time in many a drawing room and pub, and everyone wanted to look at the new planet. King George III of England appointed Herschel as his Court Astronomer. In this position, Herschel was able to devote himself entirely to astronomy. In 1787 he discovered two moons of Uranus. This enabled the mass of the planet to be calculated. It is 14½ times that of the earth.

116 / ASTRONOMY

11-5 Uranus with four of its five known satellites visible. The largest one, in the upper left corner, is Titania.

Uranus can often be seen without a telescope on a clear, dark night. Astronomers of Ptolemy's day and before must have seen it. However, it moves so slowly, and it is so faint, that they did not recognize it as a wanderer. In the widespread rush of interest in the new planet, old star charts were brought out, dusted off, and looked over. Uranus' position in the sky at the time a chart was made was determined. Then the charts were compared with modern ones. On twenty-three of the old charts an extra star was found at the predicted position of Uranus. The first observation had been recorded by Tycho; the remaining twenty-two were telescopic observations, the oldest made in 1690. Uranus had been seen in six different constellations of the zodiac, and mistaken for a star, during the 90 years before its recognition as a planet.

Perturbations in Uranus' Orbit

As soon as an orbit had been worked out for Uranus, tables were made to show the position of the planet in the sky in future years. Of course, the perturbing effects of the other planets were taken into account. In 1798 the observed positions were off from the predicted ones by only a small amount. This error was not serious. But by 1810 they were off more, well above the possible error of a skilled observer with a good telescope. And the errors were increasing. In 1821, a French astronomer, Alexis Bouvard, decided to rework the orbit calculations, in the hope of making more accurate tables. He had twenty-two prediscovery positions available covering 90 years, and nightly records of the planet's positions for the 40 years since 1781. To his surprise, he found that he couldn't construct an orbit that fit them all. When he included the prediscovery positions, the present position of Uranus was way off. If he used only the postdiscovery ones, the present position was off less. He decided to discard the twenty-two older observations, which were, he reasoned, more likely to be in error. He drew up his tables based on a corrected orbit.

But Uranus didn't stick to this orbit. In 11 years most of the positions in Bouvard's tables were in error. The tables had to be constantly revised, just as in the days when Ptolemy's tables were used.

If the planets were behaving according to the law of gravitation, there should have been no such differences between predicted and observed positions. It was generally believed that Newton's laws were universal—that they at least held to the limits of the solar system. If they did not, then exploration of the universe beyond Uranus—out in the realm of the fixed stars—would be difficult without his laws as guides.

Astronomers realized that the gravitational attraction of another as yet undiscovered planet beyond Uranus could be disturbing its motion, making it appear to be moving contrary to Newton's laws. In 1841, John C. Adams, a 22-year-old student at Cambridge University, decided to figure out where such a planet must be, and what orbit it would follow in order to produce the unexplained parts of Uranus' motion. Then the planet might be found in the sky.

Adams first examined the positions of Uranus from 1782 to 1822, and concluded that during this time the faster-moving Uranus was catching up with Planet X (the undiscovered planet).

118 / ASTRONOMY

As they came closer, the gravitational force between them increased, and Uranus was pulled farther from its true orbit. During this time, Uranus was speeding up—it got ahead of its predicted positions. This meant that Planet X lay somewhat ahead of Uranus and pulled it forward. From 1822 to 1842, however, Uranus' observed positions were coming closer to the predicted ones, and, in addition, the planet was slowing down. Adams concluded that during this period Uranus had overtaken Planet X and was now ahead, being pulled backward. In 1822 the difference between observed and predicted positions was greatest, so it was probable that the two planets had passed each other at that time.

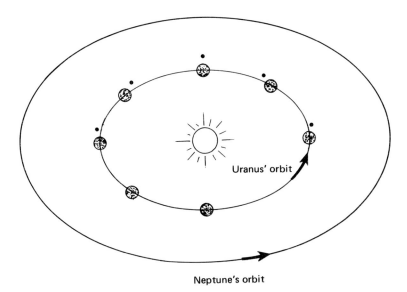

11-6 The dots represent the observed positions of Uranus while the larger circles represent the predicted positions. (The differences are exaggerated to make the diagram understandable.) This difference however, led astronomers to search for the planet Neptune.

Since the size of Uranus' orbit and its perturbations were known, Adams figured out how fast Planet X must be moving along its orbit. Then from Kepler's harmonic law, he determined the radius of the orbit.

Who Cares about Planet X?

By October 1845, Adams had a solution that satisfied him. He predicted how far Planet X's orbit was from the sun and that its mass would be about twenty-five times that of the earth. He also calculated where a planet following this orbit would be found in the sky during the next few years. He took his computations and predictions to an astronomer at Cambridge University, hoping he would look for Planet X with the university's telescope. The astronomer wasn't interested, and sent Adams to Greenwich to see the Astronomer Royal, who wasn't interested either.

Meanwhile, a well-known French mathematician somewhat older than Adams, named Leverrier, had tackled the problem without knowing of Adams' work. In the summer of 1846 he presented three papers to the French Academy on the subject. Now the Astronomer Royal pricked up his ears. He ordered the Cambridge astronomer to look for the planet in the position where Adams predicted it would be. He didn't find it. A later inspection of the two sky maps he drew showed that he had not compared them carefully enough—one star *had* moved. It was Planet X.

At the end of the summer an astronomer at the Berlin Observatory turned his telescope to the place in the sky where Planet X should be. Although no object in the telescope field showed as a disk, there were nine stars there and only eight on the chart he was using. Was one of them the planet? The next night he found that this ninth star had moved slightly with respect to the others. The new planet had been found! Adams' predicted position was almost as accurate as Leverrier's. Both men were given credit for discovering *Neptune*, as the new planet was later named.

The discovery of Neptune was another triumph for Newton's theory, and a reassurance that it could be used in further exploration of the universe.

And Yet Another

Even after the perturbing effect of Neptune was taken into account, differences remained between the predicted and observed positions of Uranus. And Neptune itself was not keeping exactly to its orbit. In 1906 an American astronomer, Percival Lowell, reasoned that there was a planet beyond Neptune. He predicted that this planet would appear small and faint, and that it would spend about the next 30 years in the constellation Gemini. Lowell searched for the planet at his Arizona observatory until his death in 1916, without success.

11-7 Neptune pictured with its two moons. The smaller one (Nereid) is shown by the arrow in the upper right corner. The larger satellite near Neptune is called Triton.

Such a faint and slow-moving object is much more difficult to find than Neptune, but by this time there were new instruments to aid in the discovery. A telescope with photographic film at the telescope focus had been developed in 1849. Time exposures could be made. The light falling on the film from a star gradually accumulates and builds up a photographic image. After photographic films were made more sensitive, it became possible to photograph objects which were 100 times too faint to be seen by looking through the telescope.

Lowell's search had been made by studying photographs. After his death, others kept on looking for the planet. The search was intensified in 1929, when a new refracting telescope was put into

URANUS, NEPTUNE, AND PLUTO / 121

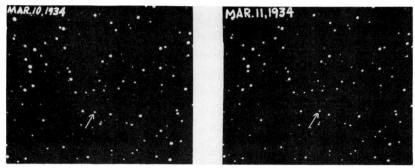

11-8 Pluto (arrow) is shown above in two positions taken in a 24-hour period.

operation at Lowell Observatory. The entire area of the sky around Gemini was photographed. A photograph of the same region was taken three times, usually within a week. Each plate recorded from 5000 to 400,000 stars.

In February 1930, Clyde Tombaugh, a 24-year-old astronomer at Lowell Observatory, was comparing photographs made on January 23 and January 29 of that year. In the area where Lowell had predicted the new planet would be found, he saw a faint object whose change of position during these 6 days appeared about right for a planet beyond Neptune's orbit. It was within 6° of Lowell's predicted position. Additional photographs were taken on several successive nights. They showed that the new planet was moving westward. Its orbit proved to be very similar to the one Lowell had predicted. Announcement of the discovery was made on March 13, the 149th anniversary of Herschel's discovery of Uranus. The planet was named *Pluto*, after the Greek god of the underworld. Its first two letters are Percival Lowell's initials. Its period is 248 years. Its mass is as yet not determined accurately.

From 1930 on, a detailed examination of photographs already made, and a new photographic survey of the region of the ecliptic has been going on at Lowell Observatory, to see whether another planet lies beyond Pluto. None has so far been found.

Now the solar system, as we know it, was complete. Data, such as distance, size, period, and mass, for all the planets is shown in Table 4 on page 124.

Test Yourself

1. Why was Herschel certain that the new "star" he had found in the constellation of Gemini was not really a star?

2. What two things showed that Herschel's discovery was not a comet?
3. If Uranus can sometimes be seen without a telescope, why hadn't it been discovered before?
4. What indicated that there might be another yet undiscovered planet beyond the orbit of Uranus?
5. When Pluto was first discovered, it was moving *westward* with respect to the stars. Does this mean that it orbits in an opposite direction from the other planets?
6. What do you think enabled astronomers to calculate that Neptune has a mass about twenty-five times that of the earth?
7. Uranus is double Saturn's distance from the sun. How many AU's is it from the sun (see Table 1, page 45). Does Kepler's harmonic law indicate that its period is 80 years?
8. **Why was it agreed that a planet beyond Uranus would move slower than Uranus?**
9. **If there is a planet** several times farther from the sun then Pluto (as **several** astronomers think, from study of comet perturbations), **why** will it be difficult to find?

Table 4. Data on the nine planets.

	Mercury	Venus	Earth	Mars	Jupiter	Saturn	Uranus	Neptune	Pluto
Average distance from sun (in AU)	0.39	0.72	1.00	1.52	5.20	9.54	19	30	40
Period	88 days	225 days	1 year	687 days	12 years	29.5 years	84 years	165 years	248 years
Diameter, in terms of earth's diameter (8000 mi)	0.37	0.96	1.00	0.52	10.9	9.1	3.7	3.5	0.5?
Number of moons observed	none	none	1	2	12	9	5	2	none
Mass, in terms of earth's mass	0.05	0.81	1.00	0.11	317	95	14.5	17.6	0.18
Weight of a person weighing 100 pounds on earth (in lbs)	39	91	100	41	267	115	106	144	?

12

The Planets

Mars has been photographed and studied more than any other planet because it can be viewed well when it comes close to the earth (as close as 35 million miles). Its surface, shown in Figure 12-1, is easily seen through a fairly small telescope. Its atmosphere is thin and rather free of clouds. The polar caps, which appear to be mantels of frost around both its north and south poles, grow larger in the winter of each hemisphere. The greater part of the planet's surface is a rocky or sandy desert, pockmarked with

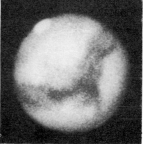

12-1 (left) The photographs of the surface of Mars was made by Mariner IV in 1965. (right) Mars, showing the polar cap and dark band around the equator. The lines on the scale bordering the left-hand photograph are about 4.3 miles apart.

125

126 / ASTRONOMY

craters. Haze and yellow clouds cross its deserts from time to time. Around the equator is a dark, blue-gray belt that changes size with the Martian seasons.

The "red spot" on Jupiter, shown in Figure 12-2, is 30,000 miles across and has been observed since about 1830, but is still not satisfactorily explained. The stripes, (light and dark bands parallel to the equator) show continuous but gradual changes. This shows that they are cloud bands, rather than the solid surface of the planet.

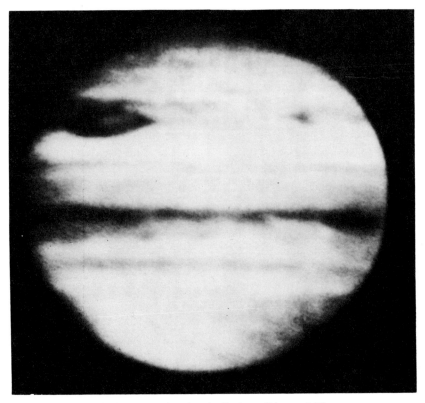

12-2 Jupiter showing the "red spot" (upper left).

The rotation period or day on a planet is given in Table 5 and can be determined by observing features such as the red spot on Jupiter and many surface features of Mars as they move across each planet's disc. All the planets except Mercury, Venus, and Pluto rotate fairly rapidly. And they all receive sunlight about half the hours of their day—their period of rotation—from 5 hours at a stretch on Jupiter and Saturn up to about 12 hours (about the same as earth's) on Mars.

Table 5. Lengths of days on the planets in terms of hours and days on earth.

	Period of rotation *(= planet's day)*
Mercury	59 days
Venus	243 days*
Earth	24 hours
Mars	24½ hours
Jupiter	10 hours
Saturn	10 hours
Uranus	11 hours
Neptune	16 hours
Pluto	6 days

*Venus rotates in the opposite direction from the other planets.

Gravity and the Atmosphere

When you throw a stone up in the air, it always falls back to earth. The upward motion that you gave it is overcome by the downward acceleration of gravity. If you throw the stone harder, so that it leaves your hand with a greater speed, it will rise farther. But the force of gravity soon slows down the stone, then stops it, then reverses its direction.

How fast would you have to throw a ball or shoot a bullet or launch a space probe so that it never came down? How fast must an upward moving molecule be going in order to escape from the earth's atmosphere or from any other planet's? If the initial upward speed given the object is great enough, it will keep on going upward, in spite of the downward pull of gravity. An object thrown upward from earth's surface must be going at 7 miles per second in order to escape.

The *escape velocity*—how fast an object would have to travel to escape the earth's gravity (or any planet's)—is the same for a molecule of oxygen, hydrogen, ammonia in a planet's atmosphere;

or for a bullet, missile, or space probe. But the escape velocities, given in Table 6, are different for each planet. They depend on its mass and the distance to its center of gravity. As you would expect, they depend on how dense and how large a planet is. While a gas molecule moving 3 miles per second can escape from Mars' atmosphere, it would have to be moving over twelve times that fast to escape from Jupiter's. This shows why Mars has a very thin atmosphere and Jupiter a very thick one.

Table 6. Masses and radii of the planets (as compared to earth) and escape velocities.

Planet	Mass	Radius	Escape velocity (mi per sec)
Mercury	0.05	0.37	2.5
Venus	0.81	0.96	6.5
Earth	1.00	1.00	7
Mars	0.11	0.52	3
Jupiter	317	10.9	37
Saturn	95	9.1	23
Uranus	14.5	3.7	13
Neptune	17.6	3.5	14
Pluto	0.18?	0.5?	1.4?

At higher temperatures, gas molecules move faster. The speeds at each temperature, however, are different for each kind of gas, but they are known. Before we can predict what gases could be present on each planet, we have to estimate the temperature of each planet's atmosphere.

Like the earth, the other planets get their light and heat from the sun. The closer a planet is to the sun, the more of the sun's energy reaches it. Measurements show that on earth the area directly under the sun and at the equator at noon (its *subsolar point*), receives about 13 calories per minute per square inch. (A *calorie* is a basic unit of heat.) Since we know how much farther (or nearer) the other planets are to the sun, we can figure how much energy their subsolar points get.

The planets shine by reflecting sunlight back out into space.

Table 7. Distances, heat received and absorbed, and measured temperatures of the planets. (Measured temperatures listed for Venus refer to the top of the cloud layer; at the surface the temperature is about 640°F.)

	Distance from sun (in AU)	Calories received each minute by 1 sq cm near subsolar point	Percent of radiation retained	Calories retained each minute by 1 sq cm near subsolar point	Measured temperatures (F)
Mercury	0.39	13.2	94	12.2	+610° to −500°
Venus	0.72	3.85	24	0.925	−27° to −45°
Earth	1.00	2.00	59	1.18	40°
Mars	1.52	0.87	85	0.74	80° to −150°
Jupiter	5.20	0.074	49	0.036	−220°
Saturn	9.54	0.022	50	0.011	−230°
Uranus	19	0.0055	33	0.0018	below −300°
Neptune	30	0.0022	38	0.00083	−350°
Pluto	40	0.00132	84?	0.0011	below −350°

This means that they are sending some of the energy they got from the sun back out into space. But not all of it. Each planet absorbs some of the sun's energy and uses it to heat itself up. Cloud-covered Venus reflects back three-fourths of the sunlight that falls on it. (You have seen clouds in our sky brilliantly reflecting the light of the sun at sunset.) On the other hand, the dull, rocky surface of Mars reflects only 15 percent back into space. Knowing how much energy from the sun each planet keeps, the temperatures can be figured out at their subsolar points. They are shown in Table 7.

Since the speeds of each molecule at various temperatures and the escape velocity on each planet are known, we can determine which gases *could* be present on the planet. They are shown in Table 8.

They could be there, but are they? We have "taken the planets' temperatures" even though the nearest thermometer is millions of miles from them. With the nearest chemistry lab equally far away, how can we determine for sure what gases are present? Astronomers did it by a study of their light.

Table 8. Gases which could be in the atmosphere of the planets, as predicted by temperature and escape velocity.

Carbon dioxide (CO_2):	Venus, Earth, Mars
Nitrogen (N_2):	All except Mercury
Oxygen (O_2):	All except Mercury
Carbon monoxide (CO):	All except Mercury
Water (H_2O):	Venus, Earth, Mars
Ammonia (NH_3):	All except Mercury and Mars
Methane (CH_4):	All except Mercury and Mars
Helium (He):	Only the outer planets
Hydrogen (H_2):	Only the outer planets

Long-Distance Chemistry

In 1666 Newton let a narrow beam of sunlight shine through a prism. What came out was not white sunlight but a band of colors. Newton saw that each color of light was coming out of the prism at a slightly different angle, thus spreading the colors into a rainbow (shown in Figure 12-3). The sequence of colors in this

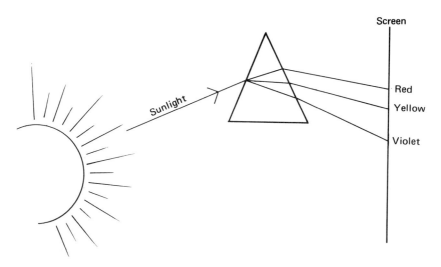

12-3 Rainbow from a glass prism, showing three of the colors.

rainbow is always the same: red, orange, yellow, green, blue, violet. Astronomers call this sequence of colors, shown in Figure 12-4, a *continuous spectrum.* Astronomers soon found that there is also light that is invisible to the human eye but which shows up on photographic film. The light that shows up beyond the red end of the spectrum is called *infrared.* The light that shows up beyond the violet end is called *ultraviolet.*

Dark Lines in the Spectrum

In the summer of 1801 a fire broke out in a looking-glass factory in Munich, Germany. The flimsy building collapsed, killing all the workers except a 14-year-old apprentice named Joseph Fraunhofer. Seriously injured, he was rescued from the flames. The ruler of Bavaria happened to be passing by and took an interest in the boy, later presenting him with 18 ducats (about $40). Young Fraunhofer used some of the money to buy a machine to grind and polish glass into prisms and lenses. Over the next few years, he spent the remainder of the money for books on spectra and light. Unwittingly, the Bavarian king had made a research grant whose dividends in scientific discovery, dollar for dollar, could well arouse envy in modern governments.

132 / ASTRONOMY

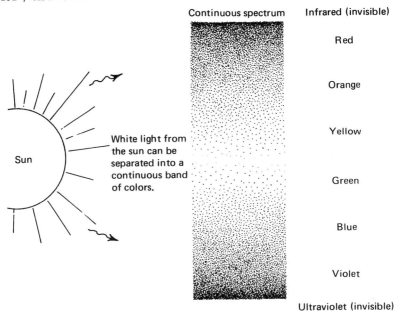

12-4 Sunlight contains light of all colors, as does the light of any star.

In 1802, an English scientist noticed several dark lines running across the continuous spectrum of sunlight. He thought these were "the natural boundaries between colors." When Fraunhofer read this, he looked for these spectrum lines. Although they are difficult to see, by 1814 he had found about 600 of them in the sun's spectrum. Whatever instrument he used, each spectrum line was always at the same place in the spectrum, forming a distinct, irregular pattern. He found that the spectra of the moon and planets also showed dark lines and that most of these were identical with those in the sun. Although he didn't believe that these dark lines were color boundaries, he didn't propose a theory to account for them.

Nevertheless, Fraunhofer's published descriptions and measurements led many scientists to look for these dark solar-spectrum lines. Among them was Gustav Kirchhoff, a professor at the University of Heidelberg, in Germany. By 1859 he had found that dark spectral lines could be produced in the spectra of artificial light sources by passing their light through various gases. Each gas gave its own pattern of dark lines—as individual as a fingerprint. Kirchhoff concluded that each gas removes certain colors from the continuous spectrum, so that these colors are missing or diminished. These missing colors produce dark lines where colors would otherwise be. He called these lines *absorption lines*, or *dark-line spectra*, shown in Figure 12-5. Today the positions of many thousands of absorption lines in the solar spectrum have been measured, and their separate patterns deciphered.

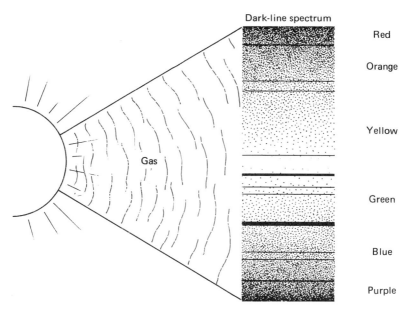

12-5 When sunlight from the sun's hot surface passes through cooler gases in the sun's atmosphere, certain colors of light are absorbed by each of the gases. If the light is then dispersed by a prism, it shows a continuous spectrum crossed by dark lines. These lines indicate the colors that were absorbed by the gases.

Fraunhofer Lines

Since Fraunhofer first discovered dark lines in the spectrum, they are sometimes also called *Fraunhofer lines*. These lines are actually breaks in a continuous spectrum. Absorption by each different cool gas (oxygen, carbon dioxide, and others) causes a different set of dark lines to appear in a continuous spectrum. Each such set of lines always has a definite pattern, so that a gas can always be recognized by its dark lines.

When hot, glowing gases send their light through a prism, we see a spectrum of bright lines of color, as Figure 12-6 shows. Each element has a particular set of bright lines—also like a fingerprint—unique and identifying. Astronomers call this a *bright-line*, or *emission spectrum*. The bright lines and dark lines of the same gas have the same patterns—the same fingerprint—whether they are in

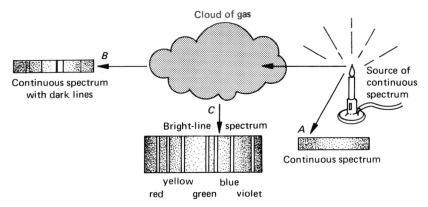

12-6 The light of a candle gives a continuous spectrum (from red to violet) as shown at *A*. If this light is passed through a cloud of cool gas (or through the atmosphere of a planet) there are gaps (dark lines) in the spectrum as at *B*. If the cloud of gas is glowing, it will form a spectrum made up of bright lines at *C*.

the visible spectrum or the invisible (ultraviolet and infrared). A dark-line spectrum of the sun is shown running down the center of Figure 12-7. The thicker dark set of bands are from the element iron. In the sun iron is not a solid, but a vapor or gas. Above and below the sun's spectrum in the figure is a bright-line spectrum of the actual element iron, for comparison. The comparison spectrum was obtained in a laboratory on earth. Notice how the bright lines coincide with the dark lines in the sun's spectrum, thus showing that the element iron is indeed present in the sun's atmosphere. Determining elements in the planets' atmospheres is done in a similar way.

As you probably know, the light by which we see each planet is reflected sunlight. If the planets were perfect mirrors and had no atmospheres, their spectra would be identical with that of the sun. But the sunlight penetrates into each planet's atmosphere before it bounces back to us. In its passage into and out of those atmos-

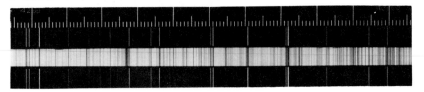

12-7 Astronomers photograph the dark-line spectrum (center) of the sun and the bright-line spectrum of the element iron (top and bottom). They can now make comparisons which indicate whether iron is present in the sun's atmosphere. What would you say?

pheres, the light picks up more dark lines in its spectrum. These dark lines show which gases are present in a planet's atmosphere. The gases thus far identified by dark lines (absorption lines) in the spectra of the planets are listed in Table 9.

Table 9. Composition of the planets' atmospheres as shown by absorption lines. (No determinations have been made for Mercury or Pluto.)

	Carbon Dioxide	Nitrogen	Oxygen	Carbon Monoxide	Water	Ammonia	Methane	Hydrogen
Venus	√	√		√	√			
Mars	√				√			
Jupiter						√	√	√
Saturn						√	√	
Uranus							√	√
Neptune							√	√

Is There Life on Other Planets?

Mars and the earth seem to be the only planets with comfortable temperatures and the lack of deadly gases like carbon monoxide, methane, etc. More important, Mars seems to be the only other planet where the temperature range would allow *organic compounds*—the material of living things—to form and endure. These complicated molecules, chiefly of carbon, hydrogen, nitrogen, and oxygen, make up all known living things—from one-celled animals to men and dinosaurs. Laboratory experiments show that these molecules are broken up, or do not even form, at high temperatures like those of the planets Mercury and Venus. They might, however, be present high in Venus' atmosphere where it is cooler.

Organic molecules will not form unless the "building blocks" of organic compounds—molecules of carbon dioxide, methane, ammonia, and certain others, are present in liquid or gaseous form, or unless there is water that can move them about. Otherwise, as solids, they cannot come together to form organic molecules. From the temperatures given in Table 7 on page 129 it is clear that there is probably no liquid water (freezing point about 32°F) on the planets beyond Mars. In addition, laboratory experiments have shown that at the low temperatures of the outer planets, many of the building-block molecules will not be gases. We cannot expect them to be present in the atmospheres of these planets.

Living things need water to exist. In addition, all forms of animal life that we know need oxygen gas to survive while plants use carbon dioxide. Mars and Venus seem the best bet among the other planets of the solar system as possible sites of life. The outer planets, Jupiter, Saturn, Uranus, and Neptune, with their methane and ammonia-laden atmospheres, look even less attractive.

But the lack of oxygen on Venus is significant. It is difficult to imagine how animal life could survive without it. Oxygen is very active chemically and is quickly removed from the earth's atmosphere by chemical combination with other elements. Oxygen in the earth's atmosphere is fairly constant only because plant life releases it continually, replacing the supply that is being used up by animals breathing and other chemical action. The lack of oxygen makes it look as if there is no plant life on Venus.

What about Mars, legendary home of a sinister and skilled race of little green men, racing through space aboard their flying saucers? The absence of oxygen on Mars makes it look as though the only possible life would be some primitive forms, like lichens.

Test Yourself

1. What is *continuous spectrum*?
2. Kirchhoff discovered the reason for *dark lines* (Fraunhofer's lines) in the spectrum. What is the reason?
3. Of what use to astronomers would *bright lines* in a spectrum be?
4. Why do you suppose that the polar caps of Mars are thought to be fairly thin coverings of frost, rather than thick ice caps like we have at the North and South Poles?
5. At the temperatures of the planets' atmospheres, would you expect to see bright lines in their spectra?
6. Why do dark spectral lines give more definite information about a planet's atmosphere than temperature and escape velocity?
7. What characteristics must a planet have in order for life to develop there, and keep on developing?
8. Why does it appear that earth is the only planet of the sun which has higher forms of life?

13

Other Members of the Solar System

In 1766, when Saturn was still the most distant planet yet discovered, a German named Johann Titius somehow hit upon a scheme for remembering the distances of the planets from the sun. He took the numbers 0, 3, 6, 12, and so on (each number after 3 is twice the one before it), added four to each, and divided by ten. The result is shown in the first seven lines of Table 10. Six years later, Johann Bode, Director of the Berlin Observatory, heard of this. He noticed the blank at 2.8 AU and organized twenty-four European astronomers into a sort of club or committee to hunt for the missing planet.

Table 10. Bode's Law is one way to remember the distances of planets from the sun.

Titius' progression (Bode's Law)	Planet	Planet's actual distance (in AU)
(0 + 4)/10 = 0.4	Mercury	0.387
(3 + 4)/10 = 0.7	Venus	0.723
(6 + 4)/10 = 1	Earth	1.000
(12 + 4)/10 = 1.6	Mars	1.524
(24 + 4)/10 = 2.8	————	————
(48 + 4)/10 = 5.2	Jupiter	5.203
(96 + 4)/10 = 10.0	Saturn	9.539
(192 + 4)/10 = 19.6	Uranus	19.191
(384 + 4)/10 = 38.8	Neptune	30.071
(768 + 4)/10 = 77.2	Pluto	39.518

As you know, Uranus was discovered in 1781, but it lay 19.2 AU from the sun. It was not the planet that Bode's men were looking for. However, it fit in well with "Bode's Law," $(192 + 4) \div 10 = 19.6$ AU) as Titius' scheme is known today, and stimulated the search for another planet at 2.8 AU, which is somewhere between the orbits of Mars and Jupiter.

The Minor Planets

A Sicilian astronomer named Guiseppe Piazzi spent New Year's night in 1801 locating and mapping each star in the constellation Taurus. He was using a newly published catalogue in which each star was listed according to its position on the celestial sphere. Searching carefully for a listed star that he could not find in the sky (later shown to be a printer's error), his attention was caught by a faint star that the catalogue failed to list.

To his surprise, it was in a slightly different position among the stars the next night. He looked at it every clear night and saw that it was moving eastward. Then in early February it reversed its direction, as a planet does. The next November, when the calculations of this wanderer's orbit were completed, it was found to lie at just about 2.8 AU from the sun. It was named Ceres (for the Roman goddess of Sicily), and joyfully hailed as the missing planet. Measurements of its size showed that Ceres is disappointingly small (only about 500 miles across). Even though its orbit lay between those of Mars and Jupiter, both easily visible to the naked eye, Ceres could only be seen with a good telescope.

Quite unexpectedly, another planet even more minor was discovered about a year later. It also circles the sun at about 2.8 AU. By 1807 two more even smaller planets had been found. By 1890 more than 300, all smaller than the first four, had been added to the list of minor planets.

The list grew rapidly after photography made their discovery easier. When a long time-exposure photograph is made with a telescope, a motor moves the telescope tube westward, following the stars as they move across the sky during the night. This makes the stars appear on the photograph as points of light (not as trails like they do when no motor drive is used). However, even with the telescope drive working, a minor planet moving in its orbit makes a short trail, like the blurred picture of a person who moved while a snapshot was being taken. Thousands of these short streaks made by minor planets have been seen on long-exposure photographs of the night sky near the ecliptic. From these, the orbits of about three thousand minor planets have been figured out. Most of them circle the sun at distances between 2.3 and 3.3 AU. There are estimates that as many as forty thousand more minor planets are waiting to be discovered.

13-1 Asteroid Icarus photographed in June, 1949. Notice the long trail.

There are only a dozen minor planets with diameters of 100 miles or more. A few hundred of them are between 25 and 100 miles across, while the vast majority have diameters of only a mile or so. Of course, there is no reason to suppose that there are no minor planets less than a mile across—we just can't see them. But even then, it is hard to see how their combined masses can be more than 1/1000 that of the earth.

Many astronomers believe that these minor planets formed from the break-up of one planet—the one Bode's committee was looking for. The irregular shapes of the minor planets do suggest that they

13-2 A meteor streaks across the sky.

are fragments from an explosion, or have been nicked by collisions, or both. As a round planet rotates, the amount of reflected sunlight that we receive from it remains the same. But the light of many minor planets varies, showing that their surfaces are irregular.

If the minor planets did at one time make up one planet, it was a very small planet indeed, unless most of its fragments have gotten away, perturbed by Jupiter, or bounced out of orbit by a collision. Some of the small fragments may have collided with the earth. What happens when a chunk of minor planet collides with the earth striking us at 8 to 10 miles per second? We see a bright "shooting star" or *meteor* streak across the sky, heated white hot by friction with the air. It might land as a *meteorite* and eventually be placed in a museum if it doesn't burn up first. Often, a large meteorite blasts a crater as it lands, as Figure 13-3 shows. Many millions of meteors burn up in the air for every one that hits the ground, and only about one per century makes a crater.

13-3 Air photograph of a meteor crater near Winslow, Arizona.

About twenty-five meteorites are found each year. Some of them are newly fallen; others lie on the ground many years before they are discovered. Most of them are stony, somewhat like rocks on earth. Others are metallic, composed chiefly of iron and nickel, and some rare ones combine both types. Meteorites were the only material from outside the earth that man ever held in his hand or analyzed in his laboratory until the Apollo landings on the moon. Even if a meteor doesn't reach the ground or is not found, its light can be photographed during the minute when it glows in the sky. A bright-line spectrum then shows some of its chemical composition.

Before a meteor enters the earth's atmosphere, it shines dimly by reflected sunlight. When it is in the earth's shadow, of course, it does not shine at all. However, while meteors are plunging through the earth's atmosphere, they give out their own light. The glossy surfaces of meteorites show that their outsides have been melted by friction with the air. Studies of meteor spectra show that part of the surface material becomes gaseous. The air nearby is heated too, and electrically charged. For a few moments there is a brilliant trail of light behind the swiftly moving meteor—mostly of glowing electrified air.

By studying the spectrum of this trail, astronomers can determine what a meteor is made of, its temporarily high temperature, and the speed and direction of its motion. From this information, astronomers can figure out the orbit that the meteor had been following before being pulled in by the earth's gravity. Many had orbits coming from outside Mars' orbit where the minor planets are. The arrival of these meteors cannot be predicted. They come in erratically and four or five can be seen from one place on a clear night, each with a different orbit.

Meteor Showers and Comets

Several times each year, on certain definite dates, there is a night or two when a great number of meteors come in—thousands in a few hours—all from the same direction. Meteors from these *showers,* as they are called, never land as meteorites. They are so small that they burn up completely in the atmosphere. At these times, the earth is crossing an orbit followed by many small particles. A year later, we cross it again, and there are meteors aplenty.

The orbits of shower meteors around the sun are not circular, but extremely long, thin ellipses with the sun at the focus near one end. In 1866 it was discovered that the meteors which shower down on us on August 11 each year, travel in the same orbit as a comet that had been seen for the second time in 1862. So it seems that shower meteors are the debris of comets—nature's "pollution" of the space around the sun. Almost every meteor-shower orbit has been found to be identical with a known comet's orbit, as shown in Table 11. Twice each year, the earth passes through the orbit of the most famous comet of all, Halley's comet, providing the May and October showers. In all other cases a comet's orbit is tilted to the earth's orbit, so that we cross it only once each year.

The main object in such an orbit is the *comet,* which is made of snow, wax, and rock fragments all frozen together. Spread out along the orbit of a comet are fragments that have come unstuck. The comet, as it moves along its orbit, seems to be falling apart, with its sandy debris trailing behind it all along the orbit, like papers in the wind behind a car with a litterbug in it. No matter where we cross the orbit there is some debris for the earth to run into, and this debris we see as meteors.

Table 11. Dates of meteor showers

Date of best display	Name	Associated comet	Period of comet (yrs)
January 3	Quadrantid	———	7.0
April 21	Lyrid	1861-I	415
May 4	Eta Aquarid	Halley	76
July 30	Delta Aquarid	———	3.6
August 11	Perseid	1862-III	105
October 9	Draconid	Giacobini-Zinner	6.6
October 20	Orionid	Halley	76
October 31	Taurid	Encke	3.3
November 14	Andromedid	Biela	6.6
November 16	Leonid	1866-I	33
December 13	Geminid	———	1.6

A meteor, whether of comet or minor-planet origin, glows as it is heated by its speed through the earth's atmosphere. But the light of a comet, far outside the earth's atmosphere, is caused by heat and light from the sun. Far from the sun, where a comet spends most of its time, all of its material is extremely cold and

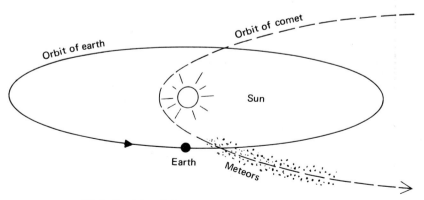

13-4 Particles of the comet spread out along its orbit.

doesn't shine. As it comes closer to the sun, we can see it faintly by reflected sunlight. It is then extremely small, a solid ball less than 200 miles in diameter. No one knows what the mass of a comet is because no comet has accelerated another body.

When a comet comes within several AU of the sun, its material begins to evaporate and these gases begin to glow, forming a cloud of light around the comet, called the *coma*. As time goes on, and the comet gets closer to the sun, the gases stream out away from the sun in a tail, as in Figure 13-5. It may be very long—hundreds of thousands of miles—depending on how much gas is boiling off,

13-5 The tail of Comet Cunningham (top) is very small, while Halley's Comet (bottom) has a long, glowing tail. Why are the streaks made by the stars longer in the top photograph than in the bottom?

and how close it gets to the sun. The coma remains as a small circle of light around the comet. The shape of the tail changes; it is almost straight as the comet moves toward the sun, curved as the comet swings around the sun in its elliptical orbit, and *precedes* the comet as it recedes from the sun. The tail always points away from the sun because the *solar wind* (thin clouds of gas moving away from the sun's surface) is blowing it outward. Changes in the solar wind have caused breaks in comet tails, some of them looking like two tails from the same comet.

Fred L. Whipple, Director of the Smithsonian Astrophysical Observatory, developed a theory of comets in the 1950's now generally accepted. He reasons that a comet is like a dirty snowball, one in which pebbles and sand are mixed with snow, wax, and other materials. When you bring such a snowball indoors where it is warm, the snow evaporates and leaves a small pile of gravel and sand. Comets do about the same, but do not stay near the sun long enough to melt completely.

When we see a comet a few million miles away, glowing in the sky night after night, we might not think of it as a dirty snowball. But in time, its orbital motion carries it so far from the sun that it cools well below freezing, and again shines only by reflected sunlight. It is again a dirty snowball. Soon it moves so far from the sun that it can no longer be seen. Then after 3 years, or 60 years, or 105 years, depending on its orbit (or how far out it goes), it returns and the spectacle is repeated.

Each time a comet passes close by the sun, it loses part of the frozen material that holds it together, just as the snowball did in the house. No wonder the comet drops fragments along its path. After several trips around the sun, we might expect so much of the "snow" to be lost that the comet could no longer hang together. Does this happen?

First, it must be said that many comets last a long, long time. Halley's comet, for instance has come back every 76 years for the last 2000 years or more. Evidently, a large icy ball doesn't lose much of its material in the few months of each pass close to the sun. Other comets are in less elongated orbits (smaller eccentricity) and never get very close to the sun. In general, comets shrink very slowly with age. One, Comet Biela, was seen to break into pieces as it passed the sun in 1846, and one group of meteor showers may have shown the death of the comet.

146 / ASTRONOMY

13-6 An artist's drawing of a meteor shower on November 16, 1833.

The shower of November 16, 1833, was a fantastic one. Many people thought that the end of the world had come. As many as 200,000 meteors could be counted from one place in a few hours. Again, on November 16, 1866, and November 16, 1899, there were striking showers, although the numbers of meteors were less each time. It seems that somewhere along that orbit there is a dense swarm of fragments and every 33 years we hit it. In the intervening years we cross the shower-meteor orbit where the fragments are spread out more thinly. It is possible that this swarm is all that is left of an old comet, with a period of 33 years.

Will there be a time when there are no more comets? Or are new comets forming today? If so, how are they formed? Where do they come from? Astronomers are attacking these problems today with the patience of Tycho and Kepler and the imagination of Galileo and Newton. But nobody yet knows the answers.

Test Yourself

1. What did Titius' formula (*"Bode's Law"*) enable astronomers to do?
2. What are the minor planets and about how many are known?
3. Where are the minor planets situated in the solar system?
4. How do astronomers suppose the minor planets originated?
5. What is the difference between a *meteor* and a *meteorite*?
6. What are *meteorites* made of?
7. Are shower *meteors* made of the same stuff?
8. What are *comets* made of?
9. What evidence is there that meteor showers are connected with comets?
10. How does a comet change while we watch it? What causes these changes?
11. Is a comet's tail always streaming out behind its direction of motion? Discuss.
12. Why do you expect a comet to "die"?
13. How do you suppose Ceres' size was measured?

14

Satellites

You probably know something about satellites. Several countries of the world send up satellites that circle the earth for a variety of reasons. They may gather information about world-wide weather, temperatures of the sea, and crops on land. Mariner IX, a space probe launched from the United States in 1971, was sent up to orbit Mars as a satellite. It took pictures of Mars and radioed them back to earth. But did you know that the moon is a satellite too? The moon is our natural satellite. A satellite is any object that travels in an orbit around a much larger body. It requires no power to keep it in orbit, where it is held by the gravitational pull of the larger body.

Our moon is large as satellites go. Although there are five larger moons in the solar system (see Table 12), they accompany planets much larger than the earth—Jupiter, Saturn, and Uranus. Relative to the size of its planet, our moon is the largest one, with a diameter over one-quarter that of the earth. Indeed, the earth-moon system, revolving around its center of mass, could be called a "double planet."

Galileo, back in 1609, concluded that the moon is much like the earth. His telescope revealed mountains, craters, valleys, and large, flat areas that he at first thought were seas. However, as our view of the moon became clearer, and after we landed on it, striking differences from the earth's landscape have become more evident. For one thing, Galileo's "seas" are dry, and most of the lunar surface is pocked with circular craters, large and small. What makes the moon so different in appearance from the other half of this "double-planet system"?

Table 12. Known satellites of the solar system

Planet	Number of known satellites	Diameter range of satellites (mi)	Ratio of diameter of largest satellite to its planet's diameter
Mercury	0	—	—
Venus	0	—	—
Earth	1	2160	0.272
Mars	2	5-10	0.002
Jupiter	12	15-3200	0.035
Saturn	10	100-2600	0.034
Uranus	5	400-1000	0.034
Neptune	2	100?-3000?	0.109?
Pluto	0	—	—

It isn't the moon's material (although the astronauts' samples show there's no water). The amount of sunlight absorbed by the moon's surface is the same as that absorbed by dark rocks or sand on earth. The spectrum of the sunlight that comes to us as moonlight resembles the spectrum of sunlight reflected from rocky material. (And, incidentally, it bears no resemblance to that reflected from green cheese!)

The reason for the moon's different landscape becomes clear if you know the moon's escape velocity and the temperature of the moon's sunlit side. The moon cannot hold an atmosphere. Three observations made before the Apollo landings show that the moon's surface is practically a vacuum. On the earth, the air scatters sunlight around into the shadow on the night side so that we have twilight for a short time after sunset. Viewed from out in space, as the astronauts have done, this shows up as a gradual shading from light to dark—the twilight zone. In Figure 14-1 however, a sharp line divides the moon's illuminated half from its dark half. Also, when the moon's orbital motion carries it between us and a star, the star's light always blinks out suddenly. If the moon had an atmosphere, the star would dim gradually as the moon's atmosphere came in front of it. In addition, all of the dark lines in the moon's spectrum are those put there by the sun's atmosphere or the earth's atmosphere.

SATELLITES / 151

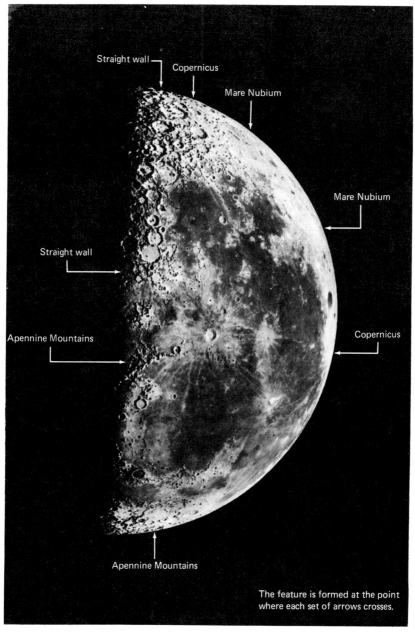

14-1 The eastern half of the moon. Notice the sharp division between the dark surface and sunlit surface.

Because there is no air on the moon, there can be no water on the moon's surface. And, without air or water there can be no wind, rain, or snow. There are no rivers to carry rock material from high places to low. No ocean waves pound a shore. On the earth, rivers, waves, wind, and glaciers are constantly removing rock material from one place and depositing it in another. They are changing the landscape and have been doing so for over a billion years. This makes the history of the earth difficult to decipher. Part of the record in the rocks is gone, and part of it is buried.

The moon doesn't have such a "poker face." The record of much that happened to it is probably still written on its surface. Because they are not eroded in the same way, similarly formed features look different on the moon than on the earth. Mountain ranges on the moon have no streams cutting into them, and are not covered by soil and vegetation. Mountains are generally higher than those on the earth.

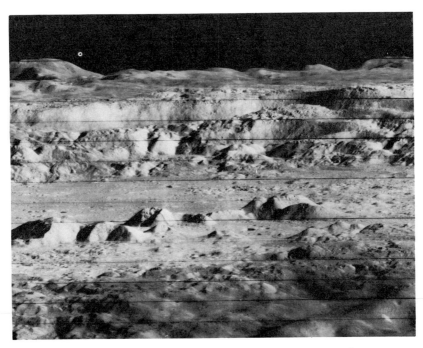

14-2 This is part of the first close-up photograph of the moon's crater Copernicus. It was taken on November 22, 1966.

The features that really set the moon's landscape apart from the earth's, however, are thousands of craters, ranging from a little more than a foot to 150 miles across. The large crater named for Copernicus shows prominently in Figure 14-2, and is a detailed view of its interior. The inside walls of large craters like this rise to as much as 10,000 feet above the crater floors and many have mountain peaks at their centers. Some scientists speculate that they are volcanoes. Many of these features do resemble Crater Lake in Oregon, an extinct volcano. Most geologists and astronomers agree, however, that most of the moon's craters were formed by explosions following the impact of fast-moving meteoroids of all sizes.

Why should the moon's surface have so many craters, while less than forty have been found on the earth's surface? On our planet, erosion would immediately begin to remove or bury a crater; all but the latest ones would be destroyed or hidden. On the moon, only the impact of later meteorites could erase an older crater. Then too, we would not expect to find smaller craters on the earth because smaller meteors are burned up as they pass through our atmosphere. In addition, the atmosphere has a slowing effect on incoming meteors. The energy of a meteor landing on the moon is much greater than if the same meteor hit the earth. Therefore, it can make a bigger hole as it hits.

Both Mars and the moon, with thin atmospheres and little or no water, have crater-pocked surfaces. Meteorites would be a definite hazard to life there. Meteorites are also the chief means of erosion on Mars and the moon. On the earth, however, a meteorite is so rare that those few that are found are treasured by museums, and meteorite erosion plays almost no part in sculpturing the earth's surface.

Many craters dot the surfaces of the *maria*—the large, dark plains of the moon that Galileo at first mistook for seas. One of these, Mare Imbrium, shows clearly in Figure 14-3. These plains may be vast outpourings of lava, but certain observations show that they are not surfaced with solid rock. As the moon enters the earth's shadow during an eclipse, the temperature drops. This very rapid drop in temperature can be accounted for partly by the absence of the blanketing effect of an atmosphere. The rapid cooling of the moon's surface also suggests that it is mostly covered with a

14-3 This view is of the edge of the Mare Imbrium and the crater Archimedes, located in the north central part of the moon.

layer of fine dust. Photographs like the one in Figure 14-4, taken in 1967 by NASA's Lunar Lander, confirmed this. It seemed likely that the dust is volcanic ash or tiny fragments of meteorites, or both.

Men on the Moon

On July 21, 1969, men explored the lunar landscape that Galileo's telescope had first revealed. Neil Armstrong and Edwin Aldrin landed the Apollo 11 spacecraft in the "Sea of Tranquillity" and left their footprints in the lunar dust. They collected and brought back samples of it. Some of the dust particles proved to be angular rock fragments but, surprisingly, over half the particles were glass "beads," spheres of glass less than one-eighth inch in diameter. These beads are thought to be spatterings of meteorite-pelted rock that hardened as they were whirled high over the lunar surface. Since then, other Apollo astronauts have collected many rock

14-4 Note the depression made in the lunar surface by one of the footpads of Surveyor I. What kind of a material might make such a depression?

specimens—coarse- and fine-grained rocks which resemble terrestrial ones that have hardened from molten lava. The Apollo 12 astronauts, who landed in the "Sea of Storms" on November 19, 1969, found larger rocks made up of angular, pebble-sized fragments of lava-like rock, bound together by glass and similar to the ones in Figure 14-5. The astronauts of Apollo 14, 15, and 16 found much the same surface material. While drilling several holes, they encountered a firmer soil about 4 feet down.

It will take years to learn all that these rocks and dust can tell us about the moon. How much of the "lava" came from volcanic outpourings is not certain. Some came from rocks melted by the terrific impact of meteors. Studies have shown that one specimen of lava-type rock brought back by the Apollo 12 astronauts is 4.6 billion years old. This great age suggests that it may be an unaltered piece of the moon's original crust.

 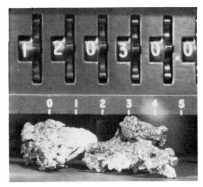

14-5 The lunar material on the left is much larger than that on the right. Both are made of lava-like material held together by glass.

The astronauts found the moon's surface bleak and lifeless, pockmarked by craters of all sizes, strewn with boulders large and small, and covered with dust. They found no evidence of living things, and did not expect to. They were able to exist on the moon only because of the life-support systems brought from earth, attached as back packs to their space suits. During the moon's 2-week-long day, the bright side heats to above 212°F, the boiling point of water on earth. On the night side the temperature is down to −280°F. There is no air to breathe nor water to drink.

Artificial Satellites

The earth may have captured chunks of rock too small to be seen as they orbit the earth, but until October 4, 1957, the moon was earth's only known satellite. Then the first man-made moon was launched; now there are over 1000 of them in orbit. Astronomers welcome the information radioed back by these satellites and by space probes. They cannot get such data with earth-based telescopes because the earth's atmosphere filters out much of the light coming in from the stars, and smears photographs made with the light that does get through so they show no fine details. The Orbiting Astronomical Observatory, launched into orbit in 1967, is as great an improvement over ground-based telescopes as Galileo's telescope was over naked-eye observations.

Satellites, probes, and space travel would not be possible if astronomers had not worked out the geography and movements of the solar system—the distances of the planets, their locations at any given second, their masses, and the conditions on each one. They have provided the Space Age with maps, timetables, and tourist information. Newton's laws assured men that a space ship could be put into orbit. When men organized their engineering skills for the task, it was done.

14-6 The Orbiting Astronomical Observatory gives astronomers much more information than telescopes on earth. Workmen assemble the Orbiting Observatory in the larger photograph. The smaller drawing is an artist's conception of the Observatory actually in orbit.

Putting a Satellite into Orbit

Back in Chapter 8 we followed Newton's line of reasoning that if a stone could be given a great enough forward speed it would completely encircle the earth, always falling toward, but never reaching, the earth. It would be in orbit, like the moon, on a curved path around the earth. The moon had somehow gotten a sufficiently large sideways speed. The problem was to give enough sideways speed to a satellite.

In the 1930's, American and European engineers developed accurately guided rockets. Some were used by the German armed forces in World War II. These rockets were able to push bombs in the right direction toward their target, and to keep on pushing for many miles until the fuel gave out. You can see the advantage over a gun or cannon which pushes its bullet for only a few feet

158 / ASTRONOMY

down the barrel. The acceleration of a missile continues so long that tremendous speeds can be reached. The early rockets pushed missiles for hundreds of miles before gravity brought them down. Later missiles went so fast that they could be put into orbit. The Russians were the first to do this with Sputnik on October 4, 1957.

Satellite Speed and Escape Velocity

The farther out a satellite is, the less is the pull of gravity on it. Therefore, satellites with larger orbits need less horizontal speed to keep them from falling back to earth. The early satellites orbited at distances of a few hundred miles above the earth, where forward speeds of about 18,000 miles per hour were necessary to keep them in orbit. Their periods were about 100 minutes. Later satellites were shot higher before being pushed sideways into orbits, and their periods are longer. Most of the early satellites were close enough and large enough to be seen from the ground as they moved in their orbits across the background of the stars. If you live in Boston, New York City, Philadelphia, or Washington, D.C., you can phone "Dial-A-Satellite" to find out if any satellites are visible tonight from your area.

In the zone where most of the satellites travel, the gas molecules are about one inch apart (as compared to one-billionth of an inch apart at sea level). Nevertheless, the resistance of even this extremely thin air eventually slows a satellite down. Then the pull of gravity is able to draw it nearer to the earth. As it spirals down to a level where the atmosphere is more dense, friction with the air heats it up and tears it apart. The satellite becomes a meteor. Many of the early satellites "died" like this after 2 or 3 years. Figure 14-7 shows Explorer I, the second U.S. satellite, which is in an elliptical orbit out to 4000 miles from the earth, and is expected to have a lifetime of 200 years.

Space probes, which do not orbit the earth, must be designed and fueled to go faster than 25,000 miles an hour, the earth's escape velocity. Like other members of the solar system, they travel in orbits around the sun. In a sense, they are new (artificial) minor planets. Among these are the Mariner Venus probe, launched in 1962, and Mariner IV, a Mars probe that took the photograph in Figure 12-1, page 125. Various Lunar Orbiter satellites have been

14-7 The Explorer I satellite.

guided into an orbit around the moon. These are satellites of a satellite, something not yet found occurring naturally in the solar system.

Test Yourself

1. What, in simple terms, is a *satellite*? (Your answer should apply to the moon as well as to man-made satellites, without getting complicated.) Is the earth a satellite of the sun?
2. An airliner flying a repeated pattern around an airport, waiting for its turn to land, is nowadays said to be orbiting the airport. Is this a correct use of the word "orbiting?" Explain.
3. What, in simple terms, is a *crater*?
4. How do geologists and astronomers believe that most of the moon's craters were formed?
5. What are the moon's *maria*?
6. What is a *space probe*?
7. From what he saw in his telescope, Galileo concluded that the moon was much like the earth. In what main ways was he right? In what main ways was he wrong?
8. Scientists predicted that astronauts would walk on a thick layer of dust when they landed on the moon. On what grounds did they make this prediction?
9. If a space probe is launched faster than 25,000 miles an hour, it cannot orbit the earth. It has reached "escape velocity" and must fly away. But does it fly off in a straight line and eventually leave the solar system, or does something else happen? If so, what? If an orbiting weather satellite can be called an artificial moon, what can a space probe be called?

15

The Sun

Photographs of the sun made through a telescope look like Figure 7-8 on page 77. They show a smooth, round disk, called the photosphere (meaning sphere of light), marred only by sunspots. The photosphere looks hard and solid. However, it is known that the temperature of the sun's surface is about 11,000°F. With such a high temperature the sun's material must be entirely gaseous rather than solid or liquid.

How did astronomers find out the temperature of the sun's surface, 93 million miles away? Glowing objects like the sun give out light in all colors of the spectrum. But they do not give out equal amounts of light in all colors. Around the turn of the century, two German physicists, from laboratory experiments, discovered that the hotter the object the greater part of its light is given out nearer the violet end of the spectrum. They were able to tell the temperature of a glowing object by the color of light that it gave out with greatest intensity. For the sun it is in the yellow-green part of the spectrum, and so the sun has a yellowish tinge in the sky. Stars cooler than the sun have their maximum intensity in red light and look reddish. Stars hotter than the sun have it in blue light and look somewhat blue in the sky.

Sunspots

Sunspots, which seem to mar the surface of the sun, are gigantic areas that appear dark in comparison with the areas that surround them. Although spectra show that the gases in the sunspots are about 2700°F cooler than those surrounding them, they are still very hot. Gas in a sunspot is moving at about 2 miles per second, flowing outward and upward from the center of the spot and

162 / ASTRONOMY

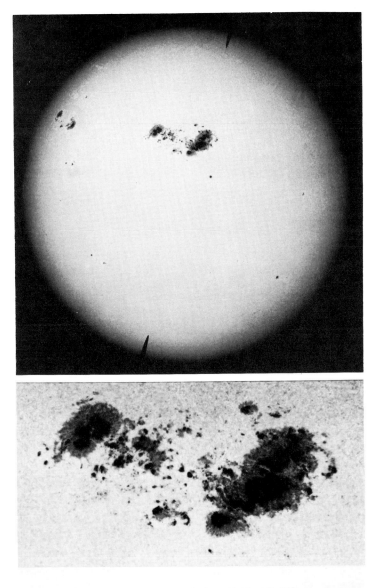

The Sunspot of April 7, 1947

15-1 The top picture shows a large sunspot. The same sunspot is enlarged in the bottom photograph.

downward and inward near its edges. Sunspots vary from specks 500 miles across to more than 50,000 miles across. Astronomers have found that sunspots appear in large groups about every 11 years. Over a hundred sunspots, for instance, were seen in 1948 and in 1959, while very few were observed in 1954 and 1965. No one has yet explained this 11-year cycle. It has long been known however, that when there are many sunspots, radio and telegraph communication on earth can be disrupted. Also, the earth's weather shows certain changes that seem to follow the sunspot cycles.

Aside from the sunspots, the photosphere appears smooth and featureless to the unaided eye. However, as we can see from Figure 15-2, the photosphere really has a mottled surface resembling rice grains. The smallest "grains" are about 300 miles across. The individual grains are columns of hotter gases, rising from layers deeper below the sun's surface. As the rising gas reaches the top of

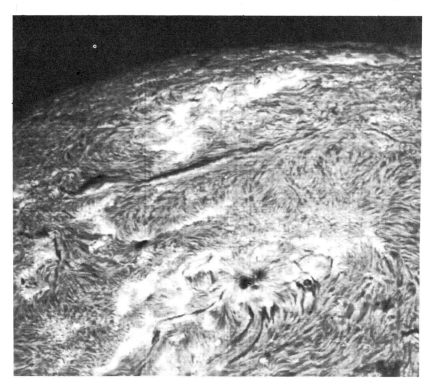

15-2 The picture is a closeup of the sun's surface which resembles many grains of rice.

the photosphere, it spreads out and sinks down again. The darker boundaries of the grains are the cooled gases sinking back into the photosphere. The whole surface of the sun appears to be active. Gases are always violently bubbling up and sinking down.

The Chromosphere and Corona

The photosphere ends sharply. Ordinarily, there appears to be nothing beyond it. However, in the seventeenth century, during an eclipse of the sun, several observers first described a narrow red streak or fringe around the edge of the sun. This was seen just an instant before and after the sun was completely hidden. It is the sun's atmosphere, just above the photosphere. It is called the *chromosphere* (which means sphere of color), and could only be seen when the bright photosphere was covered by the moon. Today, special instruments can photograph the chromosphere at any time.

Violent storms in the chromosphere send up fingers of glowing gas, called *prominences*, shown in Figure 15-3. They arch over great distances, as much as 100,000 miles above the surface of the sun, at speeds of 100 miles per second or more. In other prominences, material can be observed rapidly moving downward, back into the sun. Astronomers are not sure what causes prominences.

15-3 A view of a solar prominence taken by Donald H. Menzel.

The chromosphere extends only about 5000 miles above the photosphere. It merges into the outermost part of the sun's atmosphere, the *corona*, shown in Figure 15-4. Like the chromosphere, the corona was first observed only during total eclipses. Because the corona extends out farther than the chromosphere it can be seen when the chromosphere and the photosphere are both covered by the moon. Like the chromosphere, it is usually invisible because the photosphere is so bright. The outer part of

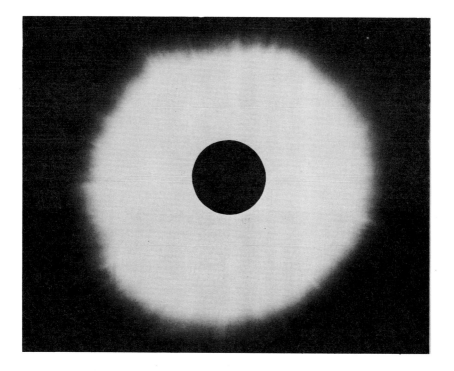

15-4 The solar corona, photographed by Georges Van Biesbroeck in Brazil at the eclipse of May 20, 1947.

the corona shines by sunlight reflected from tiny particles of dust. Its inner part, however, shines because it is so hot. Astronomers have measured the temperature of this inner part at almost 2 million degrees Fahrenheit! It seems odd, indeed, that the outermost atmosphere of the sun—so near the extreme cold of outer space—should be its hottest part. Here is something for astronomers and physicists of the future to explain!

The Sun as a Source of Energy

Each second a vast amount of energy is pouring out from the photosphere. If the sun's interior were not producing energy this fast, the photosphere would cool off. Studies of the earth show that it has been receiving about the same amount of heat and light from the sun for the past 4½ billion years.

How does the sun continue to produce so much heat and light for billions of years? The earliest answer to this question was simple, but wrong. Until about 1800, men thought that the sun's material was burning, the way wood burns in a fireplace. The rapid union of oxygen and another substance, the carbon in the wood, gives off heat and light and produces carbon dioxide. Combinations of other elements also produce heat and light. But eventually all the wood is gone and the fire goes out. The fireplace is empty except for a pile of ashes that will not burn. The sun, shining for so long, would have consumed all its materials long before now if it were simply on fire.

In the 1850's, two scientists, one English and one German, presented another idea. The sun shines, they said, because it is shrinking. As it shrinks, its particles fall inward. The gravitational force of the sun is so strong that shrinking 150 feet per year could provide all the energy given off. This is a reasonable theory until you look at it more carefully. In the 350 years that man has been observing the sun with telescopes, this would amount to a 2-mile change in diameter—much too small to be measured. Going backward, however, we find that 20 million years ago the sun would have been twice as large as it is now. In rocks laid down at that time on earth are the bones of elephants, horses, and apes. They could scarcely have survived this tremendous heat nor could their bones have been covered by sand and silt carried by liquid water.

It wasn't until this century that scientists came to understand how the sun produces energy. It involves processes that do not occur naturally on the earth, although scientists can nowadays make them occur. Since the sun is very large, its force of gravity is extremely strong. Deep inside the sun, the effect of gravity results in tremendously high temperatures and pressures. In such conditions, hydrogen is turned into helium. In a hydrogen-to-helium reaction, four hydrogen atoms are crushed together and joined to make one helium atom. You would expect this helium atom to weigh exactly four times as much as one hydrogen atom.

However, the helium atom has slightly less than four times the mass of a hydrogen atom, so some of the mass, or matter, is missing. *This matter has been converted into energy*, as Figure 15-5 shows. Scientists call this kind of change (as from hydrogen to helium) a *nuclear reaction*. Some matter is changed into energy or some into matter, in nuclear reactions. Inside this sun the newly created energy moves outward and finally appears at the surface as light and heat.

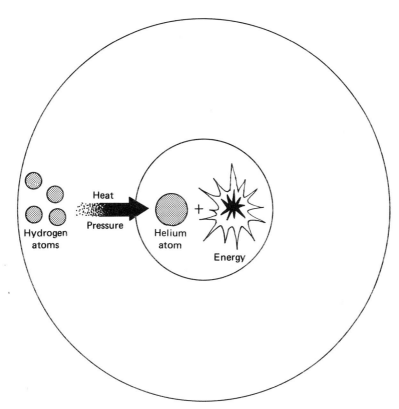

15-5 Hydrogen in the sun, subjected to a great deal of heat and pressure, turns into helium and releases tremendous amounts of energy.

The amount of hydrogen in the sun is known. The temperature and density of each zone in the sun has also been figured out. Thus, the rate of the hydrogen–helium reaction in the sun's interior can be calculated. The calculated rate of energy release is very close to the amount of energy given off by the sun each second.

Therefore, scientists believe that this is the chief nuclear reaction going on in the sun. Other nuclear reactions may add or subtract small amounts of energy.

You may be asking yourself, "But how long will the sun keep producing energy?" Well, only about 50 percent of the hydrogen deep in the sun is hot enough and dense enough to permit the nuclear reaction to go on. Knowing the mass of the sun and the amount of hydrogen used each year, scientists have figured out that the sun could shine for another 66 billion years.

However, long before the hydrogen is all gone, the sun will change, partly because the rate of reaction will drop. Still, nothing much could change for another 16 billion years.

It seems likely that the other stars also produce their energy by nuclear reactions. We would expect these other, more distant suns, to be made up of hot gases too. If the temperature, composition, and size of a star are about the same as those of the sun, we could expect that the hydrogen–helium reaction would be operating. If the conditions of other stars are quite different, then other nuclear reactions may power their furnaces. To find out, we must learn more about the other stars.

Test Yourself

1. What is a *sunspot*?
2. What causes the grainy appearance of the sun's photosphere?
3. Name two features of the sun that are invisible except during eclipses. Why can't you see them at other times?
4. What is the sunspot cycle?
5. About how long has the sun been producing light and heat at its present rate? About how long will it continue in the future?
6. Why can't the sun be shining by burning its material like wood in a fire?
7. If the sun is not burning in the usual sense, what is the source of its energy?

16

The Stars as Other Suns

Long before astronomers had telescopes or had worked out the temperature of the sun, or what makes it shine, they guessed that the sun is a star. On a clear dark night you can see about two thousand stars. None of them look anything like the sun. Nevertheless, Copernicus and Tycho reasoned that the sun would look like a star from far enough away.

The stars are so far away that distance alone would make them look different from the sun—points of light rather than a brilliant disk. The first three stars whose parallaxes were measured are up to 1½ million miles farther away than the sun. Only about seven hundred stars have parallaxes large enough to be measured even by large modern telescopes—their distances are up to 4 million times that of the sun. The millions of faint stars visible through the largest telescopes are even farther away than that.

Light from the sun, traveling at 186,000 miles per second, takes 8 minutes to reach us. The light that reaches your eyes tonight from Sirius left that star 8 years ago. You are not seeing Sirius as it looks tonight, but as it looked then. This distance of Sirius from the earth is often given as 8 light years. Although it sounds like a unit of time, the light year is a unit of *distance*. One *light year* is the distance light travels in a year, about 30 million seconds. So a light year is 186,000 × 30 million = 5½ million-million miles.

If the stars were all just like the sun, we would expect that the nearer a star is the brighter it would look. Vega is so bright that it is called "Queen of the Summer Sky." Sirius is the brightest star visible from the United States. Yet a star called 61 Cygni, nearer than Sirius or Vega, is so dim that it can be seen only through a telescope! This suggests that the stars are not all alike. Some must radiate more energy than others. If we could place Sirius and 61

Cygni at the same distance from us, Sirius would be much the brighter of the two. Would the sun and Vega appear equally bright at the same distance? Do the stars differ greatly in their luminosity or light-giving power?

In order to be sure, astronomers first measured the brightness of stars whose distances they knew. By the time that the first stellar parallax was measured, the brightness of many stars had been determined. If you look at the stars, you will see that they differ greatly in brightness. You can estimate by eye that one star is about twice as bright as another. Figure 16-1 shows how it can be done on photographs. Back in the second century B.C., the Greek scientist Hipparchus listed about a thousand stars according to their brightness, and divided them into six *magnitudes*. In the first magnitude he placed the brightest stars. The faintest that he could see (1/100th as bright) were in the sixth magnitude. (Note that larger magnitude means *fainter* star.) The other stars were assigned to magnitude 2, 3, 4, and 5, according to their brightness. Since the time of Hipparchus, more precise ways of measuring brightness have been developed, but the idea of a star's magnitude is still used.

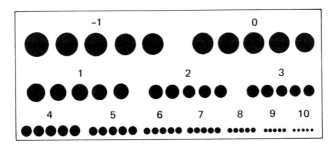

16-1 Astronomers use the above scale to estimate the magnitude of stars in photographs. Although all stars appear to us as points of light, the brighter stars appear as larger spots than the dimmer ones on any one photograph.

The *brightness* of a star means how bright it looks to us on earth. Once the brightnesses of many stars had been measured precisely, and their distances measured by parallax, it became evident that brightness does not entirely depend on distance: some stars that are farther away look brighter than some closer to us. Stars differ in their *luminosity* or the amount of light they give off. A star's brightness depends on both its distance and its luminosity.

If both the brightness and the distance of a star are known precisely, its luminosity can be calculated. It has been found that stars vary widely in luminosity. This could mean that stars are different sizes because at equal temperatures a larger object gives off more heat and light than a small one. Or it could mean that stars are all the same size but some are hotter than others. The higher the temperature, the more light and heat is given off by glowing objects of the same size. Or it could be a combination of both size and temperature that causes the differences in luminosity.

The Temperatures of Stars

Since the stars appear only as points of light in even the largest telescope, we cannot measure a star's size as we did the sun's—using distance and angular size in the sky. (The different size spots on the photograph in Figure 16-1 have nothing to do with a star's size, they only measure its brightness.)

So let's see if a star's temperature has anything to do with its brightness. On a clear night you can easily see that the stars differ slightly in color: Capella is yellowish, Orion's Betelgeuse has an orange tint, and Vega is bluish white. As you read in Chapter 12, starlight is a mix of colors that can be spread out in a continuous spectrum. These three stars send out light which includes all visible colors plus invisible radiation that photographic film can record. But they don't send out the same amount of light in each color. Capella's light is strongest in the yellow part of the spectrum. Betelgeuse sends out the most light in the orange part, and Vega's light is strongest in the blue part. This is why they appear to have different tints to their light.

Laboratory experiments have shown that the color of greatest intensity depends on the temperature of the glowing object that is giving out the light. You have probably watched a piece of metal being heated up. First it will glow red. As it gets hotter, it will glow yellow, then white. This also holds true for stars. Red stars have the lowest temperature, bluish white stars have the highest. Yellow stars are in between.

The temperatures indicated by these *peak intensities* have been determined in the laboratory for every point along the spectrum. So we can "take a star's temperature" even though we are many light years away from it. (See Table 13.)

172 / ASTRONOMY

Table 13. What is the relationship between the temperature of a star and its color?

Star	Approximate temperature	Color	Color (Spectral Type class)
Vega	22,050°F	blue-white	O blue
Sirius	20,650°F	blue-white	B blue-white
Canopus	15,850°F	white	A white
Procyon	12,650°F	yellow-white	F white-yellow
Capella	11,050°F	yellowish	G yellow
The Sun	10,950°F	yellowish	K orange
Arcturus	8,450°F	orange	M red
Antares	5,850°F	reddish	

The H-R Diagram

In 1911, a Danish astronomer, E. Hertzsprung, compared the colors and luminosities of stars within several distant *star clusters*. A star cluster is a large group of stars held together by gravitational attraction. Figure 16-2 shows the distribution of color and brightness in a star cluster. These clusters are so far away that their

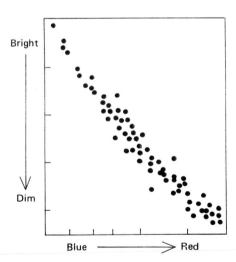

16-2 Hertzsprung's plot of brightness (vertical) against color for stars in a cluster, all of which are about the same distance from us. The plotted points, one for each star, show that blue (B) stars are much more luminous than yellow (G) stars like the sun, and that red (M) stars are the least luminous.

distances cannot be determined by parallax. However, all the stars in one cluster are at just about the *same* distance from us. So the luminosities of the stars in one cluster are related to the brightnesses we measure. Hertzsprung plotted a graph of the brightness of each star against its color, studying many clusters but using a separate diagram for each. He found that the points (each representing one star) are not distributed over the plot at random. They do not show all combinations of brightness and color. Instead, most of them lie in a band which slopes from highly luminous blue stars down to red stars of low luminosity. In these far-off clusters, the very luminous stars are the hot ones. With lower temperature, the luminosity decreases.

Two years later, a Princeton astronomer named Henry Norris Russell found that the nearby stars at known distances, like those in distant clusters, do not show all combinations of luminosity and temperature. The majority of them plot along a narrow band which he called the *main sequence*. Such stars are now called *main-sequence stars*. Russell's work showed that the relationship between temperature and luminosity holds for the nearby stars. Hertzsprung's showed that it holds for more distant stars. Therefore, diagrams like Figure 16-2 and 16-3 are known as Hertzsprung-Russell, or H-R, diagrams.

Figure 16-3 is a modern H-R diagram showing the work of many astronomers on thousands of stars. The main sequence slants across the center of the diagram from upper left to lower right, as in Figure 16-2. The blue stars near the top of the main sequence are very hot and very luminous, about 10,000 times more luminous than the sun. Those at the bottom are cool red stars, 1/100 as luminous as the sun.

Most stars lie on the main sequence, but not all. There are three patches of stars on the H-R diagram: the *red giants*, the *supergiants*, and the *white dwarfs*, whose combinations of temperature and luminosity place them separate from the main-sequence stars, and indicate that they are somehow different.

Although we started out to show that the sun is like the stars, it now looks as though the stars differ greatly among themselves. Yet the sun appears to be one of them. It differs from the other main-sequence stars no more than they differ from each other. There are many stars whose only difference from the sun appears to be their distance. The sun, in fact, happens to be a rather un-

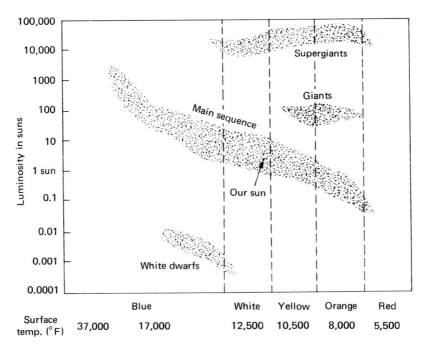

16-3 The H-R diagram. A star is represented on the diagram by a dot located, from left to right, according to its temperature; and vertically, according to its luminosity, given in terms of the sun's luminosity.

distinguished star—a main-sequence star near the middle of the sequence. It is not blue (extremely hot) and not red (cool, as stars go) but yellowish-white, of middling temperature.

The H-R diagram gave a means of estimating the distances of stars too far away to be measured by parallax. Without it we would know the distances of less than a thousand stars. Let's say we find a star whose brightness in the sky is only one-tenth of Vega's. Its spectrum, however, is like Vega's in peak color and in the pattern of bright lines. The main-sequence stars whose luminosity and temperature we could measure show similar line patterns for similar temperatures. This is double assurance that the star is very like Vega. So we can place it near Vega on the H-R diagram and simply read off its luminosity. Now that we know both its brightness and luminosity we can tell how far away it is.

We have seen that the higher the temperature of a main-sequence star, the more luminous it is. So far so good. But what causes the different temperatures of these main-sequence stars? Could it be differences in composition? Could it be that main-sequence stars are made up of different chemical elements and compounds? That the atoms and molecules in their surface layers are of different kinds in different stars?

Composition of the Stars

Astronomers have found that a yellowish star like Capella gives a spectrum very like the sun's, showing Fraunhofer lines of hydrogen, helium, calcium, iron, and most of the other atoms we know. All stellar spectra have patterns of dark and bright lines. But there are differences in the spectrum lines of different stars. Some, like Sirius, show very strong lines of hydrogen, some show the lines of helium stronger than hydrogen. Some show many lines made by molecules, others show only lines made by atoms.

These spectrum-line patterns show which elements (such as hydrogen, helium, carbon, etc.) are present as gases in the atmospheres of stars. Do the different line patterns mean that the stars are made of different materials?

It turns out that most of these differences are due to the different temperatures of the gases just above the surfaces of the stars. For instance, molecules can exist only in a fairly cool gas, atoms in a hotter one: and in extremely hot gases, atoms are broken up. And there are differences in how different elements act at different temperatures. For instance, at very high temperatures hydrogen is broken up but helium atoms stick together.

The strength (blackness on the spectrum photograph) of the various lines indicates how much of a particular molecule or atom is present. Stars of similar temperature have line patterns that are alike not only in placement on the spectrum, but also in intensity. But at the same time, by making allowances for the different form in which an element occurs at different temperatures—in molecules or as atoms—a star's spectrum does reveal the chemical composition of the star. And we find that most stars have about the same composition: 75 percent hydrogen, 20 percent helium, and 5 percent of all other elements.

So it isn't composition that determines the temperature and hence the luminosity of a main-sequence star. Could it be mass? How could we determine that mass of a star?

Pairs of Stars

Back in Ptolemy's day, anyone who wanted to be an army officer had to have good enough eyesight to see that there are two stars at the bend of the Big Dipper's handle. A short distance from the brighter star, Mizar, is another star, called Alcor, only one-fifth as bright. In 1650 an Italian astronomer, John Riccioli, took a good look at Mizar through his telescope and was amazed to see that it is not just one star. He saw that it is a pair of stars very close together. As more and more astronomers began to use telescopes, many more stars were seen to be double (see Figure 16–4). Most often, one star in the pair is considerably fainter than the other.

16-4 Photographs of a double star, Krueger 60, made in 1908, 1915, and 1920, showing the orbital motions of the two stars around each other.

Because the two stars in a pair are so close together in the sky, it is evident that they lie in almost the same direction from us. Astronomers wondered if they were stars with the same luminosity at different distances or if they were stars of different luminosity at the same distance. Because this was long before the first parallax measurement and even longer before the first H–R diagram was plotted, they couldn't tell.

Sir William Herschel and his sister Caroline found over 700 double stars with their telescope. In 1804 they were comparing old and new charts of the sky where the bright star Castor is. Their telescope showed Castor as a pair but they found that its two bright stars were in slightly different positions. So they plotted the positions of the two stars for several months. Then it was easy to see that the two stars are moving around each other like the stars in Figure 16–4. They are moving in orbits like the two masses in Figure 10–2 on page 102, or like the moon around the earth. Herschel could not tell for sure whether just one or both were in motion, but it looked as though they were both moving around their center of mass.

Newton had said that his laws are universal, but this was the first time that they could be confirmed outside the solar system. When about 45 years of measurements on another double star near the Big Dipper were plotted in 1827, it was clear that both stars in the pair are moving in ellipses about their center of mass with a period of 60 years. This double star, and many more, follow Kepler's laws, which Newton had explained. Each star is moving in an orbit, accelerated by the mass of its partner. Gravitational forces between stars are like those in the solar system but much larger. The distance of the star pair from us, and the apparent diameter of the orbit (in degrees on the sky) tell the size of the orbit in miles. The period then gives the orbital speed (distance around the orbit divided by the time it takes to go around once). The orbital speed, in turn, gives the acceleration of the orbiting star, and this gives the combined masses of the *two* stars.

If they were about the same size (moving like the masses in Figure 10-2), then the mass of each star is about half the combined mass. If one of the stars is much smaller than the other, it moves in a large orbit, barely wobbling its heavy companion, like the moon wobbles the earth. If we can measure the relative sizes of orbit and wobble, this equals the ratio of the two masses—like earth–moon = 80—so we can calculate the two individual masses. The members of a star pair are generally not too different in mass, and the ratio of the two masses can be measured from their individual orbits around their common center of mass.

Doppler Shift

Most double stars' orbits are edge-on to us (or almost so). This means that as the stars go around each other, they alternately come toward us and then away from us. If a star is moving toward us or away from us, its motion will show up in the star's spectrum. The bright or dark lines there are shifted toward the blue end of the spectrum if the star is moving toward us, and toward the red end if the star is moving away from us. This is the only way such movement can be detected, because a star coming toward us will stay in the same place on the celestial sphere. And it is so far away that even a high speed toward us would cause no change in its brightness.

Light is a wave motion. The energy of light comes to us from the distant stars (and from the nearby light bulb) in the form of waves. The wavelength (the distance from one wave crest to the next) is different for each color of light—shorter at the blue end of the spectrum and successively longer toward the red end.

In 1842 an Austrian physicist named Christian Doppler showed that if the source of the light is moving toward you the waves it is sending out get crowded, and their wavelengths are shortened slightly (Figure 16-5). In this *Doppler effect*, as it is called, each wavelength is shifted to another place in the spectrum, because its wavelength has changed. The shifts are extremely small and can be recognized only because the whole pattern of Fraunhofer lines is shifted in the same way. From the amount of Doppler shift, the speed of a star coming toward us or going away from us can be measured.

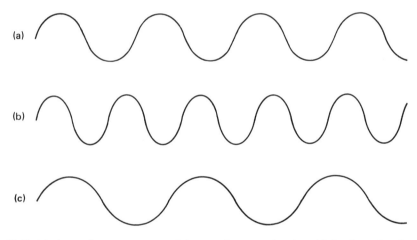

16-5 (a) shows the wave pattern of the sound sent out by the whistle of a train—the sound you would hear if you stood at a crossing near a motionless train. As the train moves toward you, the wave pattern becomes that shown in (b). And as the train passes the crossing and moves away from you, the wave pattern changes to that shown in (c). The same effect of motion on light waves changes the light's color slightly. This can be the shift of the line patterns and used to estimate a star's speed toward us

Masses of the Stars

Astronomers have found the masses of about fifty of the millions of stars in the sky—all of them double stars.

The masses of double stars are believed to be typical of stars as a whole for two reasons. A large proportion of stars are double, or triple. Almost all colors, temperatures, sizes, and luminosities are represented among double stars.

Double stars have shown us that in main-sequence stars the *luminosity increases with increasing mass.* The more luminous stars are the more massive ones. A star twice as luminous as the sun is about eleven times as massive. One four times as luminous as the sun is twenty-two times as massive. The hot blue stars at the upper end of the H–R diagram are the most massive (perhaps seventy-five times the mass of the sun). The cool red stars at the lower end of the main-sequence are least massive (perhaps one-tenth the sun's mass).

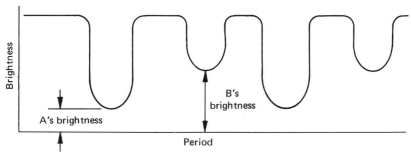

16-6 Some double stars are so close to each other that they look like one star even in a large telescope. This plot shows how an eclipsing double star "blinks" at us when star A comes between us and star B, then B eclipses A. Note that the brightness changes repeat periodically and reveal the individual brightness of each star, although we only see one image. The combined light shows both Doppler shifts, so astronomers can measure the two orbital speeds in between the eclipses.

The amount of material that a main-sequence star contains (its mass) seems to determine its temperature and therefore how luminous it is. This suggests that the stars on the main sequence are all made up of much the same material—that they have similar chemical compositions. It is the *amount* of material that seems to determine how much energy a star emits, and how hot it is.

We might also guess that the way the material is arranged is similar too, and that the nuclear processes which heat up these main-sequence stars are similar. Since our sun is one of these stars, it seems probable that its energy source, the conversion of hydrogen to helium, is theirs too.

Spectra, Photographs, Radio Waves

The spectra of stars, together with some help from Newton's laws, have told us about the temperature, luminosity, mass, and distances of the stars. Millions of light years from them, we have nevertheless used a thermometer, a measuring stick, a weighing scale, and chemical analysis on them!

You may be wondering just how astronomers photograph spectra of stars. For the last 100 years, they have used an instrument called a *spectrograph*. This instrument is a device that spreads starlight into a spectrum, and is as important in modern astronomy as the telescope. The spectrograph has contributed to almost every important discovery in astronomy in this century.

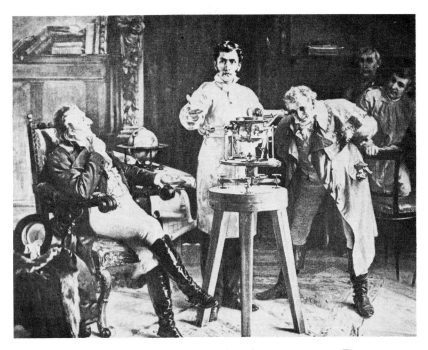

16-7 Fraunhofer demonstrating one of the first spectroscopes. The spectroscope enabled astronomers to photograph the spectra of stars, thereby learning much more about the nature of distant stars and other objects.

One last word about spectra. The invisible parts of a star's spectrum include x-rays, far beyond the violet and ultraviolet; and radio waves, far beyond the red and infrared. It may surprise you to learn that light and radio waves are all part of the spectrum energy. They behave according to the same laws. You may have heard that the speed of radio waves is the same as the speed of light. This is because radio waves are the same kind of thing as light.

Since about 1950, a new branch of astronomy has grown up, called *radio astronomy*. As you might expect, the instruments used by this branch are called radio telescopes. They consist of large, steerable antennas and sensitive radio receivers. A radio telescope acts as its own spectroscope—simply by tuning the receiver, the astronomer can find the equivalent of bright lines or dark lines. As we shall see later, important new discoveries, and new questions, have come from radio astronomy.

16-8 This radio telescope is called "Big Dish" after its dishlike shape. The newest form of radio telescope which can catch radio waves from all known radio sources in the universe, has five antennas (dishes).

As you can guess, the invention of photography was a great boon to astronomers. Today's telescopes work as giant cameras. A single photographic plate can record many thousands of star images—or many thousands of spectra. Photographic plates record colors and lines in parts of the spectrum not visible to us. By long exposures, film will capture stars too far away and too dim to be seen by the human eye even with the best of telescopes. (A photographic film can collect and store light energy—the eye cannot.) Most modern discoveries about the universe have involved rather little eye-to-telescope work—but much photography.

Test Yourself

1. How long a time is a *light year*—or is this a unit of time at all? If not, what is it?
2. What is meant by the *magnitude* of a star, first mentioned by Hipparchus and now used by most astronomers?
3. What is the relationship between *brightness* and *luminosity*?
4. What two things could account for differences in luminosity between different stars?
5. What two things determine a star's brightness or magnitude as seen from earth?
6. What, in simple terms, is the relationship between star color and temperature? (That is, what would be the main light color of a very hot star, or of a rather cool one?) Does "red hot" mean "very hot" in astronomy or in metalworking?
7. What is the relationship between color and luminosity of stars on the "main sequence" in the Hertzsprung–Russell diagram?
8. What discovery confirmed the supposition that Newton's laws applied outside the solar system?
9. What does the Doppler shift prove about light, and how is it used to learn more about stars?
10. Do main-sequence stars differ more in composition or in mass?
11. Which is more massive: a blue main-sequence star or a red one?
12. Why doesn't a Doppler shift make a blue star look red?
13. How can astronomers tell whether a star is moving toward us or not?
14. Do you have to know the distance to a star to measure its Doppler shift?

17

Lives of the Stars

Early astronomers thought of the stars as eternal and unchanging, like jewels decorating the celestial sphere. When modern astronomers realized that the stars are pouring out energy at a tremendous rate, it became clear that each star cannot shine forever and must be constantly changing. The stars are using up some of their material as they convert mass into energy. And, as hydrogen is changed into helium, a star's material changes.

So the stars should be slowly changing, or aging. However, there seems to be no evidence that they are doing so. The bright stars listed over 1800 years ago in Ptolemy's *Almagest* can still be seen as bright stars. Reddish stars that Tycho Brahe described in the sixteenth century are still reddish. No change seems to have taken place while men have been looking at the stars and recording what they saw. But, perhaps this only means that the rate of change is very slow. For instance, if we watch a person for an hour or so, we would not see much change due to his growing older.

Nevertheless, if the stars do change steadily, one result ought to be visible. Unless all stars formed at the same time and changed at the same rate, there should be stars in various *stages of development*—like babies, teenagers, and oldsters among people. Astronomers know that there are different types of stars: main-sequence stars, red giants, white dwarfs, and variable stars. Each type may represent a different stage in the life of a star, like childhood, youth, and middle age. Or perhaps they do not; maybe each different type was formed the way it is.

Most stars belong to the main-sequence group. Although it is a concentrated band on the H–R diagram, the main-sequence group includes stars that differ widely. Massive hot blue stars at the upper left are over a million times more luminous than the cool red stars with low mass at the lower right. Measures of masses of double stars show that the blue main-sequence stars have fifty times the mass of the sun, and that red main-sequence stars are one-twentieth of the sun's mass. So the main-sequence group is really a sequence from massive, luminous blue stars to red stars of little mass and low luminosity. In this sequence, two stars will be nearly the same if they plot close together on the H-R diagram. That is, if two main-sequence stars have the same color, they are nearly the same in luminosity and mass.

Most stars fit in the main sequence, but what about the others? The white dwarfs are the next most numerous. The H-R diagram (Figure 16-3, page 174) suggests that white dwarf stars are small—not only from their name. Although their temperatures are higher than that of the sun, their luminosity is about five hundred times lower. This could only mean that they are small. Each square mile of a white dwarf's surface is equally as hot as each square mile of a main-sequence star at the same temperature. So the same amount of light is being given out by a square mile on both of them. But the white dwarfs must have fewer square miles—they are smaller than main-sequence stars. This explains how a white dwarf can be less luminous than a larger main-sequence star with the same temperature.

On the other hand, the red giants are 100 times more luminous than main-sequence stars of the same temperature (below them in Figure 16-3). This suggests that they are much larger. The supergiants above them on the H-R diagram are very, very luminous stars, about 10,000 times more luminous than the sun. They must be even bigger than giant stars. Red giants and supergiants are rarer than white-dwarf stars.

Within the main-sequence group, stars of low mass and low luminosity are most numerous. Are these small, faint, red, main-sequence stars common because more of them were produced? Or are there more of them because a star stays longest in this stage of its life?

Stellar Evolution

There is one way to decide whether the different types of stars we see are stages in the life of any star. Since we know the structure and composition of the stars, as well as how they produce their energy, a theory of how a star develops can be worked out. Then the stars in the sky can be compared with the stages of development predicted by this theory. Do they fit in? Would a star be expected to shine steadily for a long time in its main-sequence stage? What type of star is oldest? Which is youngest? Early in this century, astronomers began thinking about these questions. The theory of *stellar evolution* as it is called, is still far from complete. Many of its details may change before the century is over.

As we have seen, the sun appears to be a sphere of gas held together by gravitational attraction. Nuclear reactions convert hydrogen to helium in the sun's central core, where the pressure of overlying material keeps the density and temperature high. The energy produced by the nuclear reactions in the core works its way through the outer layers of the sun until it reaches the photosphere. Then we see it as sunlight.

When the hydrogen in the core is all converted to helium, no more energy can be produced there by that nuclear reaction. Then the sun will no longer be a main-sequence star—it will change in color and luminosity so that it plots in a different region of the H–R diagram. From estimates of the amount of hydrogen that it contains, and from measurements of its luminosity (the rate at which it is producing energy), astronomers have figured out that the sun can remain a main-sequence star for about 15 billion more years. Evidence from earth's oldest rocks indicated that the sun was shining with about the same luminosity 4 billion years ago. Thus, the total time it will spend in the main-sequence stage may be about 20 billion years.

A main-sequence star that has the same surface temperature as the sun has the same luminosity, as H–R diagrams show, and thus it produces the same amount of energy per second as the sun. It also has the same mass as the sun, so it contains about the same amount of hydrogen. And like the sun, it will remain a main-sequence for about 20 billion years.

The luminosity and surface temperature of a star on the main sequence depends only on its mass. This suggests that in most other ways (chemical composition, structure, and nuclear processes) the main-sequence stars are similar. You might think that the more massive ones would remain in the main-sequence stage longer since they have more hydrogen to spend. But more massive stars are much more luminous (double the mass, eleven times the luminosity). So, in fact, they use up their hydrogen faster than smaller stars do. Then you might guess that all stars would remain in the main-sequence stage for the same length of time. But a star twice as massive as the sun is using up its energy eleven times more quickly. It will use its energy more quickly and remain in the main-sequence stage for a shorter time.

Red stars of low surface temperature (and therefore low mass) remain in the main-sequence stage longer. These stars are the most common main-sequence stars. The suggestion is strong that they are most common because they last longer than the others. They accumulate like bottle caps in a trash burner.

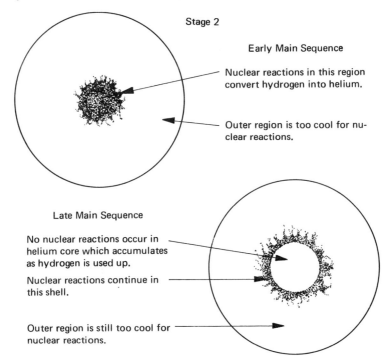

17-1 Most stars are in the main-sequence group.

Of the hydrogen used up, only 0.7 percent of the mass is converted to energy. The star does not change its mass very much during its life in the main-sequence stage. However, the chemical composition of the central part of the star gradually changes from mostly hydrogen to mostly helium. Finally, there comes a time when nuclear energy can no longer be produced by converting hydrogen to helium because the central part of the star (where conditions permit nuclear reactions) is all helium. The main-sequence stage, called stage *1* in Figure 17-1 is over.

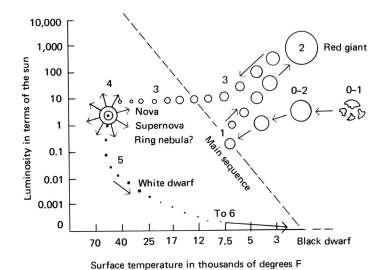

17-2 The evolution of a G-type star (like the sun) shown on the H-R diagram. Arrows do not indicate movement of the star; they show the order of its stages of evolution, as predicted by present-day theory.

But the life of the star is not over. As the hydrogen nuclear processes slow down, energy is generated by another process, which takes over in stage *2*. As the helium core begins to get cooler, outward movements of the gas and radiation pressure can no longer balance the pull of gravity toward the center. The helium core shrinks, and energy is released by this contraction. During contraction, the helium core becomes very much hotter than in stage *1*, causing large amounts of energy to flow outward through the star. This causes a rearrangement of the star's outer layers. They expand, and the star grows much larger. Like all expanding gases, these outer layers cool off. The star which was small and yellow becomes large and red. Although more energy is being radiated by the star as a whole, each square mile of its enlarged surface area is sending out less energy, as Figure 17-3 shows. The star is no longer a main-sequence star. It has become a cool but luminous *red giant*.

188 / ASTRONOMY

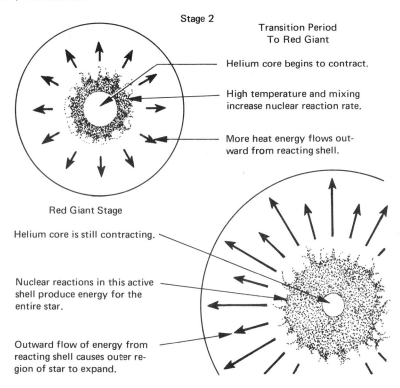

17-3 The sun will eventually become a red giant.

As contraction of the core continues, the center of the star grows even hotter. The pressure, temperature, and density there now permit the helium itself to change by new nuclear reactions into other elements. The energy produced in this way stops further contraction of the core and causes the gas there to expand again. However, the outer layers of the red giant shrink to nearly their former size. The new source of energy from helium nuclear reactions keeps the star as luminous as before, but its surface temperature rises because the star's photosphere is smaller. The star's luminosity and surface temperature change. It is no longer a red giant, and moves to stage *3* shown in Figure 17-4.

Astronomers have shown that the red-giant stage is much shorter than the main-sequence stage. A star which had been like the sun would remain a red giant for about 100 million years. A more massive star would be a red giant for a shorter time (about 1 million years). Thus, there are fewer red-giant than main-sequence stars in the sky.

LIVES OF THE STARS / 189

17-4 In stage *3* of a star's evolution you will find stars like Nova Herculis 1934 (above). Its brightness varies in a regular pattern.

A Star's Future

What happens during the remainder of a star's life is more uncertain. Many astronomers are busy today trying to figure this out. They think that at stage *3* (which may last for 1 million years) the star becomes unstable—explosions may take place inside it and it may begin to pulsate. This would make it a *variable star*, getting brighter and dimmer at regular intervals, as it expands and contracts.

Novae form in stage *4*. The word *nova* means new star and Galileo thought the one he saw was really new. But they are stars which suddenly become much brighter—mostly bright enough to be seen with the eye alone. The increased luminosity is caused by a sudden release of energy—the star blows off its outer layers. If the explosion is great enough, it will become a *supernova*. All that would be left then would be its small dense core—which is now a white-dwarf star (stage *5*).

By stages *3* and *4* (variable and nova stages), the only thing left to supply energy is the conversion of helium to other elements. When a star reaches the white-dwarf stage, all that is left is its small dense core. In that old core, no more hydrogen or helium remains to produce energy by nuclear change. White-dwarf stars are extremely dense, and their material is no longer in the form of separate atoms; it is called *degenerate matter*. They are not solid material—no known solid material is that dense. Some white dwarfs are almost a million times the density of the sun.

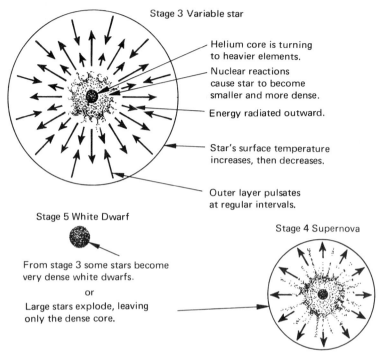

17-5 Stages 3, 4, and 5 in a star's evolution.

This degenerate material cannot produce any more energy. Slowly the old core—the white dwarf—cools off, like a poker taken out of the fire. The white-dwarf stage, like the main-sequence stage, is one in which stars linger for a long time; perhaps as long as a billion years. So we should find many white dwarfs. They should make up about 10 percent of the stars near the sun. But their low luminosity makes them hard to find. More distant ones, farther from the sun, are impossible to see even with the largest telescopes.

Astronomers can show that in a few hundred million years a white dwarf fades to 1% of the sun's luminosity. In time it will cease to shine at all, and becomes a black dwarf—a cold, nonluminous very dense sphere. It would take a white dwarf many trillions of years to reach this stage. It is possible that there are no black dwarfs among the stars because the universe may not be old enough for any star to have yet reached the final stage.

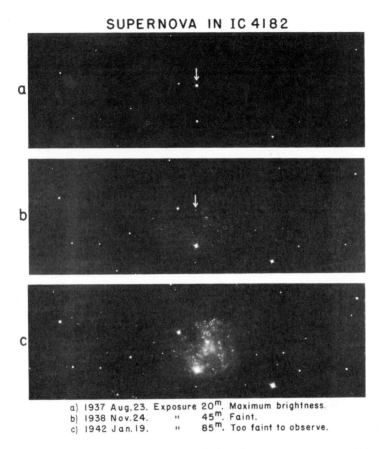

SUPERNOVA IN IC 4182

a) 1937 Aug. 23. Exposure 20^m. Maximum brightness.
b) 1938 Nov. 24.   " 45^m. Faint.
c) 1942 Jan. 19.   " 85^m. Too faint to observe.

17-6 The photographs show the evolution of a supernova over a period of 5 years.

Baby Stars

If all the stars had formed at the same time, hotter stars (like massive blue stars which use up their energy quickly) would have left the main sequence by now and become red giants. There would be no stars like them in the sky. But there are still lots of hot, main-sequence stars. If the theory of evolution of main-sequence stars is correct, this means that the stars were *not* all born at the same time. It also suggests that stars are being formed right now.

The Birth of Stars

What would a "new-born star" look like? Do stars suddenly begin their lives on the main sequence, as spheres of gas with nuclear processes going on efficiently in their cores? Astronomers do not think so. Some process has to get gaseous material together in individual spheres with high temperature and pressure in the core. Where would this material come from? It has to be gas that is not already in a star. Is there any such material available?

Not all of the material that we can see in the universe is in stars. Some of it is in *nebulae* (much thinner clouds of gas). The nebulae shown in Figure 17-7 are associated with individual stars and probably ejected from them. But there are also very much larger clouds of gas, like the one shown in Figure 17-8, glowing as they absorb and re-emit the light of stars which happen to be nearby. Some clouds do not glow and just block the light from the stars behind them. These dark nebulae look like clouds of smoke. As we will see later, they are clouds of dustlike particles and gas—a sort of smog.

In 1946, Bart J. Bok, then at Harvard College Observatory, found many such dark clouds of dust in the nebula shown in Figure 17-9. He also found smaller dark spheres of dust here and

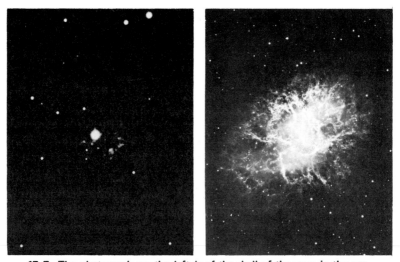

17-7 The photograph on the left is of the shell of the nova in the constellation Perseus, which brightened in 1901. This picture, taken almost 60 years later shows the star back to normal at the center of the nebula. On the right is the Crab Nebula, a large mass of gas blown out of a supernova in which a small white dwarf was left at the center.

LIVES OF THE STARS / 193

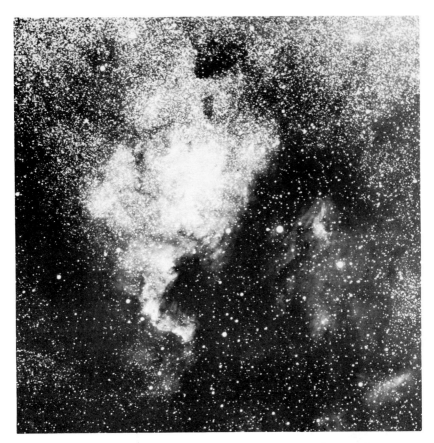

17-8 The "North-America" nebula in the constellation Cygnus, photographed with a 10-inch reflecting telescope.

there in this nebula and in others. These small globules of dust are believed to be stars starting to form. In a nebula, atoms and molecules of gas and particles of dust slowly begin to collect together, attracted by their gravitational pull on each other. This is called stage *0-1*. As one of these globules gets larger, it exerts more gravitational attraction; more and more material is attracted to it. The pull of gravity toward the center gives the collection a spherical shape. In time, gas and dust from a large region of the nebula fall into the globule and become part of it.

Slowly the globule contracts because of its own gravity. This contraction increases the pressure and temperature inside. It soon becomes so hot that the dust also becomes gaseous. Then it gets

17-9 Nebula M8 in the constellation Sagittarius. Notice the small dark spherical dust clouds, thought to be stars early in stage *0-1* of Figure 17-2.

so hot near the center that the gas gives off light, radiates, and a star is born. This is stage *0-2* shown in Figure 17-2.

In time, the contraction of the star raises its central temperature and pressure high enough so that the nuclear reaction forming helium from hydrogen can begin. Then it is a main-sequence star, about to begin its development through stages *1* to *5*. If it is massive—a big new star—it enters the main sequence at its upper left. If its mass is small, it enters at its lower right. As soon as it is born, the star's mass places it at one spot on the H–R diagram's main sequence.

During the relatively short contraction stage young stars have very low luminosities. Therefore, not many of them would be visible, even through large telescopes. Developments in stage *0–1* would take place so quickly, however, that photographs of a region like that in Figure 17-9 made in the next 10 or 15 years might reveal small changes in the globules.

Figure 17-10 shows three small clusters of very young stars, bright enough to light up the wisps of nebula which remain near them. They are probably in stage *0–2*. Photographs like this suggest that the stars in a cluster all formed at the same time. If the stars in any one cluster are all the same age (yet different in mass), a study of the stars in clusters should tell whether the theory of stellar evolution is correct. Star clusters are two types: *open clusters* and *globular clusters*. The latter are densely packed spheres of perhaps 10,000 stars.

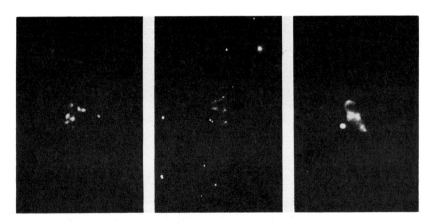

17-10 Several clusters of small bright stars associated with nebulae.

NGC 2264, shown in Figure 17-11, is a small open cluster of stars in a cloud of gas and dust. Only the more massive stars have reached the main-sequence stage. Other, less massive ones, are still in stage *0–2*.

17-11 The very young cluster NGC 2264 near the "Cone Nebula."

Now look at Figure 17-12, a diagram typical of the globular clusters. The upper part of the main sequence isn't there. The stars which once belonged there have become red giants. The main sequence has burned down farther than it has in any open cluster. The globulars must be the oldest type of cluster. There are many red giants and, in addition, stage *3* is represented. No white dwarfs have been found in globular clusters, but all these clusters are so distant that a white dwarf could not be seen.

The numbers and sorts of stars in clusters are, in fact, just what the present theory of stellar evolution predicts. But it must be

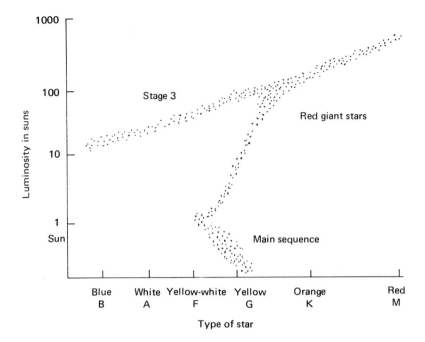

17-12 H-R diagram typical of a globular star cluster.

confessed that astronomers had one eye on the cluster stars as they worked out their theory.

It looks as though no star less massive than the sun has yet had time to evolve away from the main sequence. This means that the universe is not infinitely old. On the contrary, it appears that the universe is less than 20 billion years old.

Our Sun's Future

Figure 17-2 is a prediction of our sun's future. The sun will leave the main-sequence stage and evolve to a red giant in some 15 billion years. Then its photosphere will surely reach beyond the orbit of Mercury and perhaps even to the earth. Its luminosity will increase a hundred times. The oceans of the earth will boil away, and the baked surface will become as lifeless as the moon's. Eventually the sun will leave the red-giant stage. Perhaps it will go through a stage of variability, and explode as a supernova. Finally, it will become a white dwarf, as small as one of its planets. By then, the main-sequence stars that are today less luminous and

198 / ASTRONOMY

redder than the sun may still be shining. But a new generation of stars will have reached the main-sequence stage.

Life will be gone from the earth 15 or 20 billion years from now, but will it be gone from the universe? Sometime in the past history of our sun, probably before it reached the main-sequence stage, it acquired a family of planets. The stars show such similarity in their life stories that it would be odd indeed if our sun were the only star to have planets. In fact, if the nebular material contracting to form a star were rotating slightly, it would flatten into a spinning disk with a denser "hub" near the center. Chunks of the disk could become the planets, revolving around the hub (sun) in almost the same plane, and in the same direction, just as in our solar system.

Until about 1950, many astronomers thought our solar system was a freak, caused by another star sideswiping the sun and pulling material out of it to become the planets. But a recent theory shows that planets would form from a spinning disk as Figure 17-13 shows. Since it is likely that many globules would have

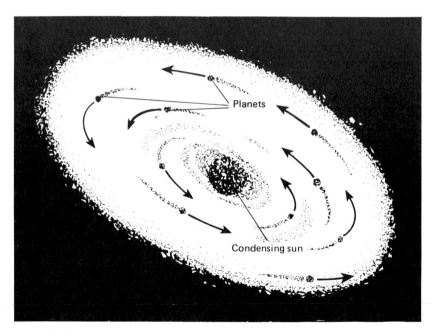

17-13 The most recent theory of the origin of our solar system suggests it formed from a spinning disk of gas and dust as shown above.

some slight rotation (and therefore contract into spinning disks) there probably are planets revolving around many stars. In some of these systems there could be a planet at the proper distance from its sun, and the proper size to have oceans and an atmosphere like ours on the earth. If so, life could (and probably would) develop on this planet. There may be millions of them, so the universe may always have living things on the planets of other stars, even after life is a thing of the past on the earth.

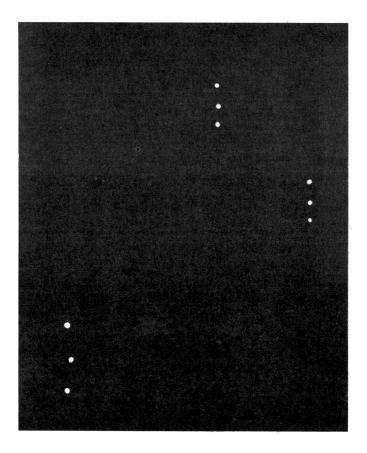

17-14 Possible evidence of a planet. Three photographs of Barnard's star and two other stars were taken at 6-month intervals. Then the three photographs were printed on the same plate, offset so that there would be three images of each star. Barnard's star is at the lower left. Notice that the middle image of Barnard's star has shifted to the right of the other two images. This is the wobble probably caused by an unseen, Jupiter-like companion.

We could not expect to see any of these planets. They could not be much larger than Jupiter. Even at the distance of the nearest stars, a non-luminous body of Jupiter's size could not be seen. However, in 1963, Peter van de Kamp of Swarthmore College discovered that Barnard's star (the second nearest star to the earth, shown in Figure 17-14) has an invisible companion. The star's very slight change of position with respect to other stars showed that it was wobbling in a small orbit (like the sun wobbles as Jupiter goes around it). He figured out that the mass of this invisible companion must have to affect Barnard's star in this way. He found it to be very much like Jupiter in size. It is not massive enough to be luminous and so it may well be the first planet of another star to be discovered.

Test Yourself

1. What is happening within stars to make astronomers think that they are constantly changing and cannot shine forever?
2. About how long is the sun expected to stay about as it is?
3. Why are main-sequence stars that are more massive than the sun expected to remain in the main sequence for a shorter time?
4. When the main-sequence life of a star (stage *1*) has ended, what has happened within the star?
5. What is the main source of energy in new stars just forming from dust and gas?
6. What is happening to the helium in the core of a red-giant star while it is a red giant, or in the pulsating-variable stage that follows?
7. By what process does a red giant become a white dwarf?
8. A "black dwarf" star is a theoretical possibility: a dead, cold, small and compact heavy sphere that cannot be seen. Why do astronomers think there may be no black dwarfs anywhere?
9. For what reason is the study of new stars in a cluster a good test of the idea of stellar evolution?
10. What evidence is there that the universe is less than 20 billion years old?

11. How could Peter van de Kamp know that Barnard's star has a companion when he couldn't see the companion?
12. If you were to see a bright star suddenly appear in the sky tonight, which you recognize as a nova, would you say that it exploded today? A million years ago? What would you need to measure in order to say when it exploded?

18

Our Galaxy, the Milky Way

Everyone who has looked at the stars, far from city lights, on a clear, moonless night has seen the Milky Way, crossing the darker sky like a faintly glowing ribbon. This band of milky light lies in a ring around the celestial sphere.

The Milky Way looks different from the rest of the sky, and Galileo found out why. His telescope showed it to be made up of stars that are much fainter and much closer together than those in other parts of the sky. That is why they all blur together when you look at them. But why should there be so many stars and such faint ones in this particular band around the celestial sphere?

In 1750 Thomas Wright, an English telescope maker, saw that if all the stars were spread out in a thin disk, with our sun at the center of the disk, the Milky Way would be explained. Standing at the center of this disklike universe (shown in Figure 18-1) we would see fewest stars along the directions toward A or A'. Along directions B or B', however, the disk extends for a greater distance. Therefore, if we look toward B or B', we would see many more stars. Most of them would be farther away than any seen along directions A or A', so more of them would look very faint. Wright's idea was that when we look along the B or B' directions, (parallel to the surface of the disk) we see the Milky Way. And since the stars are in a thin disk, we see it as a ring extending completely around the celestial sphere, as shown in the lower drawing of Figure 18-1, looking down on the disk from the top.

Some 30 years later, Sir William Herschel decided to test Wright's idea by counting the number of stars in many areas of the sky, and estimating the distance to the farthest star that he could see in each area. This he did by assuming that the distant stars are equally

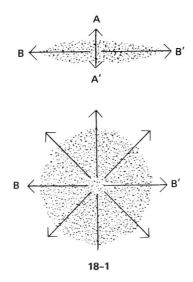

18-1

luminous, then measuring their brightness. By 1785 he had studied almost 700 different regions. In some of them, the field of view of his telescope showed only a single star, in others as many as 600, and the relative numbers agreed with Wright's prediction. Herschel's star counts and distance measures showed that the sun is inside a great disk-shaped system of stars, which came to be called the *Galaxy* (from the Greek word for milk).

How Big Is the Milky Way Galaxy?

Figure 18-2 is one of Herschel's sketches based on this long study. It is a cross-section through the Galaxy. An observer on earth, at the center of the disk-shaped Galaxy, sees the Milky Way as he looks around him (left to right on the diagram). Herschel's observations seem to show that the Milky Way has very irregular edges. He figured out its diameter and thickness. He also saw that large, glowing nebulae (clouds of dust and gas), lie in the Milky Way and help to produce its soft and beautiful light.

Herschel thought that there was a long deep cut or *rift* into the margin of the Galaxy. Figure 18-2 shows the rift as a deep indentation on the right side of this cross-section. The smaller indentations on the left edge indicate other smaller rifts. Figure 18-3 shows the appearance of the sky in the direction of the rift. This rift runs lengthwise along the Milky Way for some distance beyond

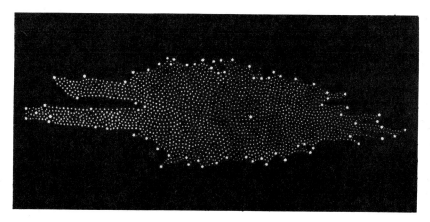

18-2 Herschel's cross-section through the Milky Way.

the edges of the photograph and cuts the Milky Way in two. It does look as though the Galaxy ends just beyond the foreground stars in the dark rift. And, up to less than a century ago, astronomers thought that these rifts were like cracks in the Milky Way, through which we could see out into the dark empty space outside the Galaxy.

But these dark areas can't be empty space because they aren't absolutely black. Even the darkest of them produces a little light—enough to be measured with modern equipment. They turn out to be clouds of dust particles that cut off our view of the stars beyond like a dark curtain would. They can't be clouds of cool gas because gas is transparent unless it is extremely dense. But a cloud of smoke (very small dust particles) is very efficient in absorbing light. A small amount of smoke can easily cut off the light of the stars beyond. Each particle absorbs some of the light falling on it. The particles also reflect part of the light, scattering it in all direction. Some of the reflected light gives a faint glow to the dark cloud, especially if stars are close to it.

The Dust of the Galaxy

The dark area in the constellation Cygnus, shown in Figure 18-3, is one of these dust clouds, or dark nebulae. It cuts off the light of all the stars beyond it. Because it is nearer than the other dark nebulae, it cuts off the light of more stars than they do, and caused

206 / ASTRONOMY

18-3 A portion of the Milky Way in the constellation Cygnus, showing the dark rift.

Herschel to think that the Galaxy doesn't extend very far outward there. Another dark nebula is shown in Figure 18-4. The small dark clouds silhouetted against the bright nebula in the "Horsehead" (near the newborn stars) are also dark nebulae. Like the bright gaseous nebula of Figure 18-3, these clouds of "smoke" are found in the Milky Way. It is not surprising that most of the highly luminous young stars of the main sequence are found here too. These dark clouds are the birthplace of stars.

Fifty years ago, astronomers were puzzled by the spectra of certain distant stars. According to lines in their spectra, these stars are hot, and should appear bluish to us, but these distant stars appear red. The reddening of the light of these distant stars means that there is dust spread thinly between the stars throughout the Galaxy. You can check such reddening yourself by looking at the moon or distant lights through smoke or smog. It is caused by extremely fine smoke particles that block blue light, but have less effect on red light. This is because different light colors have different wavelengths. Red light's longer wavelengths can get past small particles; blue light's short waves get scattered by them.

18-4 The "horsehead" nebula in Orion. The head of the "horse" is a dark dust cloud with bright edges.

The farther a star's light travels, the more dust it meets and the more it is dimmed and reddened. The effect of the dust is not apparent in the light of fairly nearby stars. Astronomers can figure out the amount that a distant star's light is reddened and then estimate how far the star is from us. Of course, to do this they must assume that interstellar dust is spread uniformly through space.

In laboratory experiments, the amount of dust needed to produce the observed reddening and dimming in stars has been worked out. Throughout the Galaxy, between the stars and nebulae, there are about a hundred tiny dust particles in each cubic mile. Interstellar space is not really crowded. The dark nebulae, which cut out the light almost completely, may contain only a hundred times more particles in the same amount of space.

This dust we have been describing is probably composed of specks of carbon and frozen compounds of hydrogen, oxygen, and carbon, such as ice, carbon dioxide, and methane. In addition to dust there are also gases (chiefly hydrogen and helium) in the interstellar material. Estimates of the amount of gas and of the stars' masses show that the material of the Galaxy near us is about equally divided between stars, on the one hand, and nebulae and interstellar material on the other.

During the nineteenth century, astronomers realized that the Milky Way Galaxy did not end abruptly at its border as Figure 18-1 shows it to. The stars are spread throughout the Galaxy and just thin out gradually near the edges. They tried to place the edge of the Galaxy by finding where the stars begin to thin out. Bigger and better telescopes allowed them to see farther and farther out, but interstellar dust (then unrecognized) makes distant stars seem much fainter, and therefore farther away than they are. Since the dust effect increases with increasing distance, it makes the Galaxy's stars appear to thin out much nearer to us than they actually do. Until early in this century, the edge of the Galaxy was estimated to be 10,000 light years away from us. When the dimming by interstellar dust was taken into account, this estimate increased to 20,000 light years.

These measurements were taken in the general direction of the constellation Auriga. Because astronomers thought that the sun was at the center of the Galaxy, they believed that the distance to the opposite edge of the Galaxy (in the direction of the constellation Sagittarius) would also be 20,000 light years. This would make the Galaxy's diameter 40,000 light years. In the direction of Sagittarius the stars are more thickly crowded (Figure 18-5) and it was difficult to measure how they thin out.

Is the Sun at the Center?

The fact that the distant stars of the Milky Way are most crowded in the direction of Sagittarius might suggest that the Galaxy extends farther in that direction and that the sun is *not* at the center. Perhaps it did suggest this to some astronomers. But it must have appeared unlikely, just as the idea that the earth might

18-5 The Milky Way around the constellation Sagittarius.

not be at the center of the universe appeared unlikely 400 years earlier.

In 1917, Harlow Shapley, then at Mount Wilson Observatory, found many globular clusters of stars in the sky near Sagittarius. In a few of the nearer clusters he could see pulsating variable stars that he thought were a type called cepheids. From their periods he determined their luminosities, and from their brightnesses he found their distances. This gave him the distance to a few of the globular clusters. He observed the total brightness of each of these clusters and then calculated its total luminosity. It turned out to be about the same for all of them—200,000 times the sun's luminosity. It seemed reasonable that the luminosity of the other, more distant clusters would be the same. So he measured the brightness of each of the other clusters and calculated their distances. He also carefully measured their positions on the celestial sphere (their directions from us) and made a plot of all the globular clusters, each at its measured distance and direction.

This plot or map showed that they were all in a giant "bubble." Shapley reasoned that this bubble of globular clusters marked the center of our Galaxy. The sun was not at the center—it was between the bubble and the Galaxy edge near Auriga!

At the time Shapley made his measurements, the dimming effect of interstellar dust on the light of distant objects was not known. Also, it was later found that the variable stars on which he based his measurements were not ordinary cepheids. They are two or three times less luminous than he thought, and therefore not as far away. He overestimated the distance to the center, but his idea was right—the sun is not at the center of the Galaxy.

Our distance from the Galaxy center turns out to be 30,000 light years. Our distance to the edge is 20,000 light years. So the

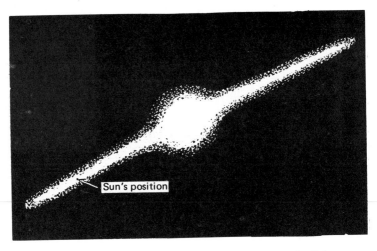

18-6 Shapley's "bubble" of globular clusters showed that the Galaxy was not a flat disk and suggested that the sun is not at the center of the galaxy.

half-diameter (radius) of the Galaxy is 50,000 light years and the Galaxy is about 100,000 light years across.

That our sun is not at the center of the Galaxy was a startling idea. Is there any evidence to back it up?

Our Galaxy Must Be Revolving

Newton's laws tell us that if the planets were not moving, they would fall into the sun. Because they are moving, they revolve around the sun, always falling toward it, but never falling into it. If the stars and nebulae and interstellar matter of the Galaxy were motionless, they would all fall together into one big mass of gas at the center. But the stars are billions of years old and have not fallen together. Therefore, they must be revolving, like a gigantic solar system. Men had always thought of them as the fixed stars, but when Herschel showed that these stars are gathered together in a disk, it became clear that they must all be moving.

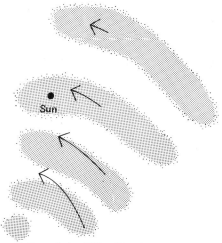

18-7 Measuring the rotation of the Milky Way Galaxy. The longer the arrows, the faster the stars are moving. Is the sun passing or being passed by the stars outside it? Inside it? On a line with it?

The center of this orbital motion is the center of the Galaxy, for the same reason that all the particles which make up the earth are pulled toward the center of the earth, as if all the earth's mass were concentrated there. If the center of these star orbits is in the center of the bubble of globular clusters, then Shapley is right. But, can the fixed stars be moving like this?

Long before Shapley's work, it was known that the stars are not exactly "fixed." Careful measurements show that many stars (the nearer ones) move a tiny bit each year relative to more distant stars near them in the sky. At first glance, these *proper motions*, as they are called, seem to go in all different directions—as random as ants crawling around on a window pane.

18-8 These photographs were taken 22 years apart and show the *proper motion* of Barnard's star during that time. In Figure 17-14, page 199, the proper motion of Barnard's star is recorded by its movement toward the bottom of the page, as shown by comparison with the two other stars, whose changes of position are due only to offset of the three photographs taken 6 months apart.

The Doppler shifts in stellar spectra show the other dimension of star motions—toward us, or away—and this can be measured for distant stars as well as nearby ones. They, too, seem to be a hodgepodge of different speeds between 100 miles per second approach to 100 miles per second recession.

After a few thousand of these stellar motions were measured, several astronomers started looking for systematic motions. The idea was that, due to chance attractions of other stars, the stars are like a huge flock of birds flying around us, going in all sorts of directions at all different speeds. But could the flock be moving south on the average? Or the stars going around us clockwise on the average? The first discovery (about 1870) was that nearby stars are streaming past the sun—or that the sun is moving past the average nearby star—at about 12 miles per second toward the constellation Hercules. This turned out to be just the sun's peculiar motion—like one bird in the flock. In order to answer the question about star orbits around the center of the Galaxy, we have to relate the average motions of stars to that center.

Orbital speed of stars in the Galaxy are similar to those of the planets. Just as Venus goes around the sun faster than the earth, so the stars nearer the center of the Galaxy should be moving faster than the sun and the other stars near it. Just as Mars goes slower than the earth, stars farther from the galactic center should be moving slower. Stars at the same distance from the center should be following the sun, or preceding the sun, at the same orbital speed, on the average.

Average Doppler shifts and proper motions show that the sun and nearby stars are going at least 200 miles per second at a distance of about 30,000 light years from the center. Not only was Shapley right, but we can use these results on star orbits to measure the mass of the Milky Way Galaxy. Although each star has its own peculiar motion like a bird in a flock, the average motion of stars near the sun will take us around the center once every 200 million years. This falling toward the center, 30,000 light years away, indicates a mass inside our orbit of about 100 billion sun masses. This is the total of all the globular star clusters, other stars, dirt, and gas between us and the Galaxy center. Of course, there is more material outside the sun's orbit, not included in this estimate, but the total mass of the Milky Way Galaxy is probably about 150 billion times the sun's mass. Although part of this

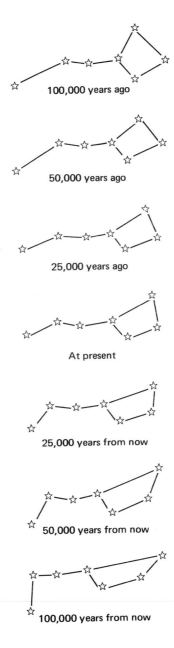

18-9 The Big Dipper through the ages, showing the effect of the different proper motion of each of its stars.

material is interstellar gas and dust and nebulae, it looks as though the Galaxy contains over 100 billion stars!

Until about 1925 it seemed likely that the proportions of the different types of stars are the same everywhere in the disk of the Galaxy outside its central bubble of globular clusters. Then Walter

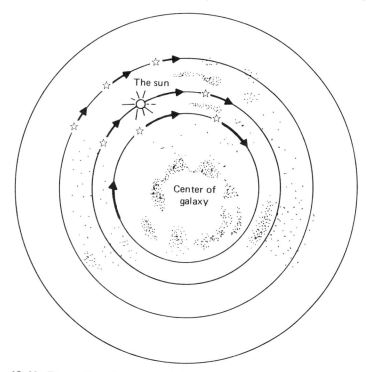

18-10 The motion of our sun and eight nearby stars as they move around the center of our Galaxy and viewed from above.

Baade of Mount Wilson Observatory showed that young, luminous, massive stars are found only in the central plane of the Galaxy (line *B–B* in Figure 18-1). They are found only in the Milky Way. And even here they are not uniformly distributed. Careful measures of their distances showed that these young stars are in three bands or *arms*—parts of circular rings in the plane of the Milky Way. The other types of stars are not confined to the arms, although they are most thickly distributed there.

Within these bright arms, stars are being formed. The material (nebulae) is there and the young stars found there show that this material has been used within the last few million years (recently, as time goes in the Galaxy). Older stars—less massive main-sequence ones, red giants, variable stars, novae, and white dwarfs are found all through the disk of the Galaxy. If they, too, were born in the Galaxy's arms, then these arms were not in the same places as present-day ones. How do arms form? How do they disappear? No one knows—yet.

18-11 The Milky Way Galaxy as it would appear to an observer outside.

From our location inside the Galaxy, astronomers have learned its shape, its size, its mass, its age, its structure, and the number of stars it contains. The sky they have studied is the interior of the Galaxy—a majestic sight on a clear, dark, moonless night. But how magnificent our Milky Way Galaxy must look from the outside, with the outlines of Figure 18-11 traced not with ink on paper but in the gleam of billions of stars and glowing nebulae.

Test Yourself

1. What does the Milky Way turn out to be?
2. What did Wright and Herschel find to be the general shape of the Milky Way, and how have modern astronomers modified that idea? That is, what is the modern idea of its shape?

3. What is the first and simplest reason why the *rifts* that Herschel found in the Milky Way cannot be empty space?
4. Is space really empty anywhere in the Milky Way Galaxy, or is there some kind of matter? If so, what?
5. Is any of the matter in the Milky Way Galaxy not in the form of stars?
6. If the sun is not at the center of the Galaxy, where is the center?
7. Shapley's estimates of the size of the Galaxy were correct in principle and yet wrong in final results. What two things threw his calculations off?
8. What is the first and simplest reason that astronomers thought the Galaxy must be revolving?
9. Why must *proper motion* (across the sky) and motion toward or away from us (motion *into* or *out of* the sky) be measured in such different ways?
10. At the center of the Galaxy are all the globular clusters. Why does this suggest that the Galaxy's central bubble is its oldest part?

19

Beyond the Milky Way

One of Wright's ideas of 175 years ago touched off an argument that wasn't settled until 1924. He believed that our Galaxy is not the only island in the sea of space. He said that there are many others and that he could point to them in the sky. This possibility was more startling and difficult to accept than his other suggestion—that the stars are gathered into a disk-shaped system. It was also much more difficult to prove or disprove.

Here and there among the stars, small faint blurs of light can be seen. Each one looks larger than a star, like a dim, roundish cloud. Beginning with Galileo, telescopes had shown that some of them are clusters of stars. As more observations were made with larger telescopes, more and more of these clouds were discovered. Many proved to be star clusters. The majority, however, still appeared as roundish or oval clouds in which no individual stars could be seen. Wright believed that these were other universes of stars, like our Galaxy, but far beyond its borders. We see them between the local stars, he said, much as the outline of a distant forest can be glimpsed between nearby trees. However, astronomers weren't convinced that they are really distant forests. It seemed more likely that they were merely thickets of nearby bushes, as the globular clusters had proved to be.

A German philosopher, Immanuel Kant, independently arrived at Wright's conclusion 5 years after he did. Kant was famous and respected, and his presentation was so clear and sensible that astronomers considered the possibility seriously and tried to find evidence that would settle the matter.

In 1781 a French comet hunter named Charles Messier prepared a list of fifty-seven star clusters and forty-five "clouds" so that they would not be mistaken for comets. They are known today by their numbers in his catalogue. The Hercules star cluster, for instance, is called M13 (it is the thirteenth object on his list). Other astronomers found many more clouds and a few more star clusters. With characteristic energy, Sir William Herschel, aided by his son John, added 4630 more. In 1864, Sir John published the complete list as the *New General Catalogue*. The objects he listed there, and those added later, are known today by their numbers in this catalogue. For instance, M13 is also called NGC 6205.

Clouds or Galaxies?

All these observations failed to determine the nature of these "clouds." Sir John Herschel included both star clusters and clouds in his catalogue because it seemed likely that, as stronger telescopes were built, all of the clouds would prove to be clusters. Sketches made in the first half of the nineteenth century showed that telescopic observations of the clouds had no more detail than does Figure 19-1. In addition, there was no way to measure their distances. Their apparent lack of parallax merely meant that they are very far away.

19-1 The larger picture of the constellation Orion was taken by Hans Pfleumer of North Brunswick, New Jersey, with his 6-inch Tessar telescope, exposure time was 10 minutes. The smaller picture shows the bright blur in Orion's sword (M42, NGC 1976) photographed through the same telescope with a 54-minute exposure time.

By this time, large instruments, like the 40-inch lens telescope at Yerkes Observatory in Wisconsin and the 100-inch mirror telescope at Mount Wilson Observatory in California, were in operation. Long time-exposure photographs made with these instruments not only revealed the beauty of these objects, but clearly showed details of structure only glimpsed at before. M31, shown in Figure 19-2, like many of the others, looks like a glowing pinwheel of light—it has spiral arms. Because of this, the clouds now began to be called spirals.

19-2 Galaxy M31, in the constellation Andromeda. It can be seen in a clear, dark sky as a faint, fuzzy patch of light, and is shown on many maps of the constellation.

Fresh from your study of the last chapter, you probably have already noticed the strong resemblance between M31 and our Galaxy. M31 looks flat and disklike. There are spiral arms, and there is a nucleus at the center, resembling the nucleus which Shapley found in the Milky Way Galaxy. However, unlike Wright's round "grindstone," M31 is oval in shape. It could be that it is tilted and that others look round because they lie at right angles to our line of sight. We see still others, like the one in Figure 19-3, in profile view.

19-3 NGC 4594, spiral galaxy, in the constellation Virgo.

However, another line of evidence soon showed that many of the unexplained clouds are not star clusters. Five years before the NGC catalogue came out, Gustav Kirchhoff's laboratory experiments had shown that a cloud of glowing gas produces a spectrum of bright emission lines and no continuous spectrum. The year that the catalogue was published, spectra of the clouds began to be photographed. Unlike the star clusters (whose spectra resemble those of the sun and stars), the spectra of clouds contain only bright lines. These particular clouds really are clouds of glowing gas. They are nebulae like the larger ones in the Milky Way, whose structure had already revealed their nature. It takes a star to make a nebula glow. In the sky near these nebulae, there are stars whose brightness makes it clear that they are part of our Galaxy. Thus, another larger group of the clouds was removed from the list of possible distant galaxies or universes.

Nevertheless, even with the nebulae weeded out, a majority of the clouds were left unexplained. Early in this century, it became clear that these remaining clouds had several things in common. For one thing, their light showed a continuous spectrum crossed by dark lines—not just a spectrum of bright lines like the nebulae. This showed that they contain stars as well as glowing gas.

The Geography of Galaxies

Back in 1917, Shapley was just announcing his idea of the Galaxy nucleus of globular clusters. It was not until 20 years later that there was enough information about our Galaxy to draw its profile. And it was another 25 years before even a trace of spiral arms was found in our Galaxy. The geography of the Milky Way Galaxy was pieced together from inside observations at the same time as the nature of the spirals was being investigated by observations of their outside appearance. Both studies aided each other by suggesting what to look for, but there were no definite conclusions from one study that could be used to prove the other.

Most astronomers agreed that spirals are disk-shaped systems of stars. Opinion, however, was divided as to whether they are fairly small systems within our Galaxy, or much larger ones outside it. The only way to settle the question was to find out how far away they are.

In some of the spirals a few faint individual stars could be observed. A series of photographs made at the Mount Wilson Observatory showed that several of these stars had brightened suddenly, like novae. The brightness of each of these faint novae was measured. If their luminosities were about the same as those of novae that had occurred in our Galaxy, then their measured brightness means that they are about a million light years away—well beyond the Galaxy's limits—and so are the spirals that contain them. It was quickly pointed out, however, that two much brighter novae (like the one shown in Figure 19-4) had previously been observed in spirals. In 1888 one in the Andromeda spiral, M31, had temporarily become so bright that it could be seen with a good pair of binoculars. This larger brightness seemed to reduce its distance to well within the boundaries of our Galaxy, if it was about the same luminosity as ordinary novae in our Galaxy. Much later, after the argument about the spirals was settled, these were found to be supernovae, which have a very much higher luminosity than novae. To add to the confusion, measurements showed that if the luminosity of the brightest individual stars that could be seen in certain spirals is the same as that of the most luminous stars in our Galaxy, then they are over a million light years away well outside our Galaxy!

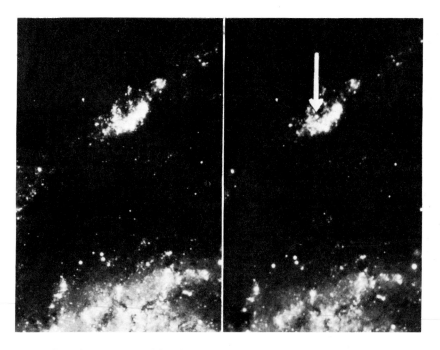

19-4 Supernova in NGC 5457. The photograph to the left was taken on June 9, 1950 (before the star flared up); that to the right was taken on February 7, 1951. Arrow points to the supernova.

And so the argument went. It climaxed in a debate in 1920 between Harlow Shapley and H. D. Curtis, an astronomer from Lick Observatory. Shapley believed that the evidence showed the spirals to be within our Galaxy. Curtis defended his belief that they are galaxies like ours.

The debate didn't settle the question, of course. It had to be settled at the telescope—and 4 years later it was. Then, Edwin Hubble of Mount Wilson Observatory discovered variable stars in three of the nearer spirals—among them M31. These variable stars had periods like cepheids. It was already known that the luminosity—the true brightness—of a cepheid is related to its period of pulsation. Comparing their luminosities (revealed by their periods) and their brightnesses, he found that these cepheids are indeed remote—far beyond the confines of our Galaxy. Those in M31 are a million light years from the sun. Thirty more spirals with recognizable cepheids were found. The cepheids showed that these spirals are between 1 and 20 million light years away from us—well beyond the edge of our Galaxy.

Now the 175-year-old search was ended; a whole new field of astronomy was opened. The frontiers of space had been extended and it was soon apparent that the universe is much larger and more complicated than anyone had supposed. Our Galaxy is not the only one. Each distant spiral is also a family of billions of stars. So our idea of "universe" had to be expanded to include the new *galaxies*, each a sister to ours. (Notice that we now distinguish ours with a capital G, for convenience.) The largest telescope in the world, the 200-inch reflector at Mount Palomar (completed in 1948), was built especially to observe galaxies. It not only can see more detail in nearby galaxies, it can also see farther than other telescopes. Galaxies extend as far as this "giant eye" can see. On photographs made with the Palomar telescope pointed away from the Milky Way, tiny images of galaxies that must be very far away are more numerous than the images of foreground stars of our Galaxy. It has been estimated from a survey of samples of the sky that there are a billion of them within the range of this telescope.

Do Galaxies Evolve?

Of course, it is clear that the galaxies that form these tiny images are very far away. But how far? How big is the known universe? If it were certain that the galaxies are all about the same

226 / ASTRONOMY

size and have about the same luminosity, then the distances could be obtained by comparing their brightness. But Hubble found that they are not all alike. He classified the different types into four main groups: *normal spiral, barred spiral, elliptical,* and *irregular*. They differ in mass, size, and number of stars, and so in luminosity. The fact that there are different types of galaxies, and that so many of them appear to grade into each other in appearance, suggests that these types may be stages in a galaxy's

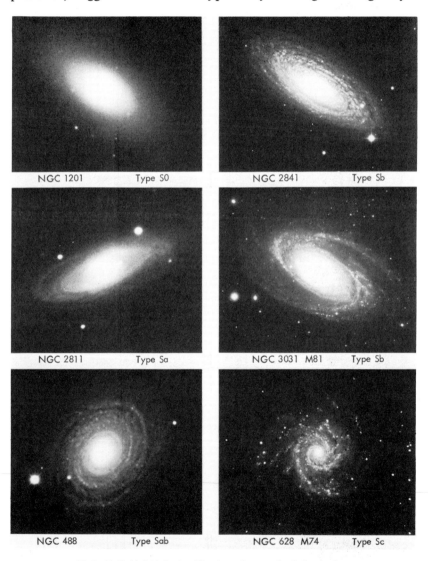

19-5 E. P. Hubble's classification of normal-spiral galaxies.

BEYOND THE MILKY WAY / 227

development, just as the different types of stars are stages in the development of a star. Thirty years ago astronomers felt sure that elliptical galaxies gradually flatten, develop spiral arms, then become spiral or barred-spiral galaxies. Later on, astronomers reversed the direction of evolution, assuming that all galaxies begin as irregulars which could be any shape. Then they evolve either through the stages of spirals (Figure 19-5), or barred spirals, (Figure 19-6), in which the arms become more tightly wound

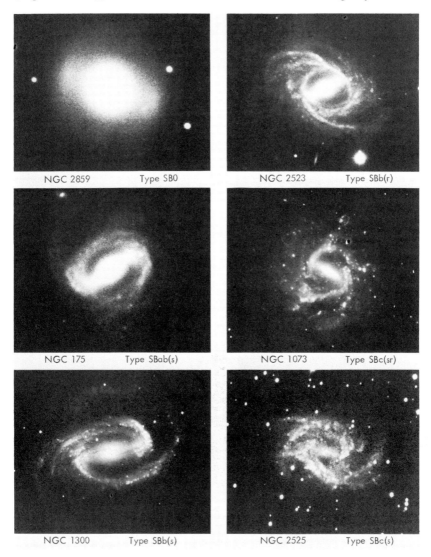

19-6 E. P. Hubble's classification of barred-spiral galaxies.

(Figure 19-7) until finally their gas and dust is completely condensed into stars and they become ellipticals (Figure 19-8).

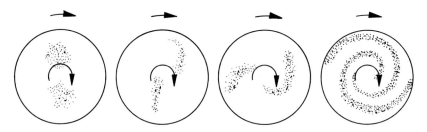

19-7 It is not surprising that gas and dust (from which new stars form) is concentrated into spiral arms. Two irregular masses of gas and dust are shown in the drawing at the left. The parts closest to the center of the rotating galaxy move fastest in this orbital motion, while those farther away trail behind.

Today, most astronomers doubt that galaxies evolve from one type to another. The fact that different galaxies are flattened by different amounts, they say, results from their different rates of rotation. The faster the rotation, the more likely the gas will become flattened into a disk and the more tightly wound its arms will become.

Astronomers have learned that galaxies form groups or clusters just as stars do. They also know that galaxies within a group or cluster move in relation to each other. But the nature of these motions remains a puzzle. If they are indeed orbital motions around some center of mass, this movement shows that the cluster of galaxies has a much larger mass than all the galaxies in the

19-8 Elliptical galaxies can be obviously an ellipse like the one in the left photograph or they may be more spherical or flattened as the ones in the right-hand drawing.

cluster. This is a puzzle. No one yet knows how these galaxies got started in motion, or whether they will remain clustered in groups forever. Perhaps the groups and clusters break up and reform, like people milling around at a party.

Test Yourself

1. Is the Milky Way Galaxy the only galaxy in the universe?
2. What is the difference between a nebula and a galaxy?
3. Why were so many of the mysterious small "clouds" eventually called "spirals" of one kind or another?
4. What causes a nebula to glow?
5. What is the important difference between the spectra of gas-cloud nebulae and the spectra of the spirals? What did it show about their compositions?
6. How did Hubble's discovery of true cepheid variables in some spirals help show that the spirals were definitely outside the Milky Way Galaxy?
7. Do astronomers think that the shape or type of a spiral represents a stage of development? Discuss.
8. In a so-called science fiction movie, a flying saucer lands on the earth, carrying invaders who say they are from another galaxy. In the course of the movie, the invaders confer by radio with their rulers back in the home galaxy. They get immediate replies—the conversations are just like our telephone conversations. Why is all this unlikely? What did the movie makers fail to understand?
9. In earlier chapters, it was stated that stars can develop and evolve. In this chapter, it is stated that galaxies probably do not evolve from one type to another. Is this a contradiction? Explain.

20

The Universe

Edwin Hubble photographed about 1300 sample regions of the sky with the 100-inch telescope at Mount Wilson Observatory, exposing each photograph for an hour. He counted about 44,000 galaxies on these photographs. They showed about the same number (several hundred galaxies) in each square degree of the sky, except in the band of the Milky Way, where dust clouds hide the galaxies behind them. He corrected the total count for the Milky Way gap and for the parts of the sky that he didn't photograph, and found that it equalled 3 million galaxies. Comparisons of their brightnesses with those of galaxies whose distances had been measured indicated that the most distant ones he could photograph are about 600 million light years away. So the estimates from the Mount Wilson photographs are estimates of the number of galaxies closer than this.

Then Hubble surveyed the same sample regions of the sky with the 200-inch telescope at Palomar Observatory, again using an exposure time of 1 hour. Again, he found an equal distribution of galaxies on every side of us. He also saw faint galaxies that didn't show up at all on the Mount Wilson photographs. The larger Palomar telescope can photograph objects four times dimmer than those which the 100-inch telescope can record—about twice as far away as the faintest ones recorded by the 100-inch.

Measurements of the size and distances of galaxies suggest that there is no edge to the universe within sight. The density of the galaxies remains the same out to a distance of 1200 million light years. Figure 20-1 will give you some idea of how enormous the universe is. Perhaps the universe has no end or border and is infinitely large. But if this were true, then there would be an infinite

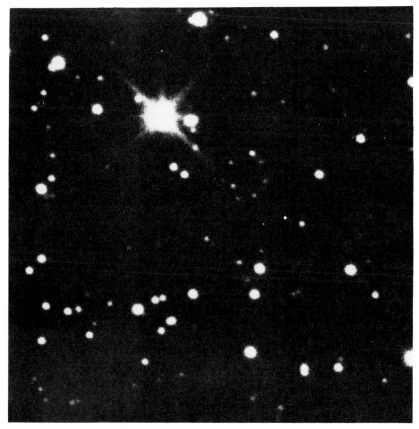

20-1 How big is the universe?

number of galaxies, giving out an infinite amount of light. How could the sky be dark at night if we are always being supplied with an infinite amount of light? Does this mean that the universe does have a border?

Movement in the Universe

Whether or not the universe is infinite, one thing we are sure of: it is not motionless. Around many stars in many galaxies, rotating planets may be revolving in orbits. Around many of these planets moons may circle. Each of the stars is moving around the center of its galaxy. Some of the galaxies are in pairs, moving around their center of mass. Others are members of clusters of galaxies, moving in complicated orbits around each other and around the cluster center.

In addition to these motions, most of the galaxies are moving *away* from us. Astronomers can determine whether a star or galaxy is moving toward or away from us by measuring the Doppler-shift of lines in its spectrum.

In the 1920's Hubble studied the size and apparent brightness of a large number of galaxies. In this way he found the distances from us of a large number of galaxies. Checking their Doppler shifts toward the red end of the spectrum, he found that the farther away a galaxy is, the greater is its "red shift." Therefore, the farther away a galaxy is, the faster it is moving away from us. This is called *Hubble's Law,* which states: The speed of a galaxy moving away from us is directly proportional to its distance away. Thus, to find the distance of a galaxy, we need only measure its speed; and to find the speed, we need only to measure the red shift.

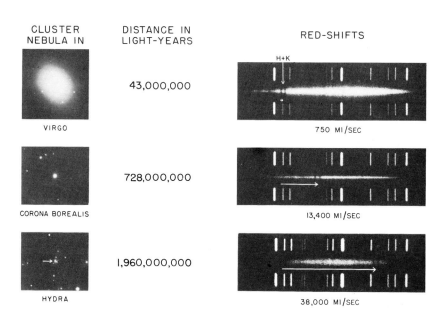

20-2 In the left column are photographs of successively more distant galaxies. In the right column is the spectrum of each of these, showing the red shift of two strong absorption lines due to ionized calcium. The arrows indicate the displacement of each line toward the red end of the spectrum (right). Below each spectrum is the speed in miles per second that each red shift indicates. The distances of the galaxies are shown in the central column.

Now, if all galaxies seem to be speeding away from our Galaxy and from each other, there is only one possible conclusion: the *universe is expanding*. Not only that, the evidence shows it has been expanding for several hundred million years. No matter how far we look, no matter how many millions of years it has taken for a distant galaxy's light to reach us, we find red-shift evidence of motion—motion that was going on when the light left the galaxy.

Hubble's Law also suggests an explanation of why the night sky is not as bright as day from the light of billions of galaxies—or perhaps an infinite number of galaxies. As the galaxies spread apart from each other, the light from any one galaxy that reaches any other is dimmed by its speed of recession. The light from each galaxy has to fill an ever-greater volume of space, and so the light energy is spread out thinner than if the galaxies were at rest.

There is one thing about Hubble's discovery that may be bothering you. Does the fact that all the galaxies are moving away from us mean that our Milky Way Galaxy is at the center of the universe? Cheated out of having our earth at the center of the universe, then of having our sun at the center of the Galaxy, are we at last to find ourselves with the dubious honor of belonging to a Galaxy so repellent that all the others are fleeing from it, so that in time the Milky Way Galaxy will be alone at the center of the universe?

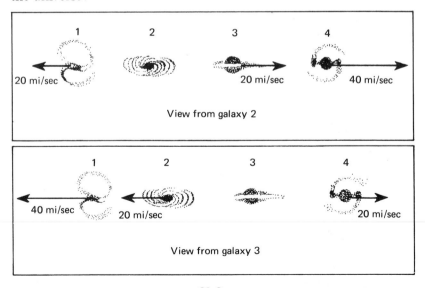

20-3

The answer to this question is shown in Figure 20-3. Let us consider four galaxies, *1,2,3,4,* equally spaced along one line. (We must also imagine other galaxies above and below this line and back of and in front of the paper, extending in all three dimensions.) The arrows in Figure 20-3 (top) illustrate Hubble's Law as seen from galaxy *2* (which we may imagine as the Milky Way Galaxy). But what does an observer on galaxy *3* see? This is shown in Figure 20-3 (bottom). He would think, of course, that his galaxy is at rest and that galaxies *2* and *4* are moving away from him on either side. He would see galaxy *1*, twice as far away, moving twice as fast as galaxy *2*. The same reasoning applies to an observer on galaxy *1* or galaxy *4*, or on all the others not shown in the figure. Unless we can tell who is moving and who is at rest, everyone gets the same view of the others. Physicists have found no way to determine which one is at rest. The movements of galaxies do not show where the center of the universe is, or if there is any center at all.

Nevertheless, an observer on any of these galaxies would see that, since all the others are moving away from him, at one time in the past they were all close to his galaxy. Since this is true for all galaxies, it must mean that in the past all galaxies were closer together than they are now. If we trace the motions of other galaxies back in time, we find that about 20 billion years ago they were all close together.

The Origin of the Universe

Thus, it looks as though all the galaxies were close together 20 billion years ago, but as Hubble showed, all are moving apart like the fragments from an exploding bomb. One major theory holds that all the material in all the galaxies was once tightly concentrated in one unstable sphere that exploded. The fragments that are moving faster—the faster galaxies—have gone farther than the slower ones. This interpretation is called the *Big-Bang* theory and is shown in Figure 20-4. The time from the initial explosion— 20 billion years—is the age of the universe. The ages of the earth's oldest rocks, the ages of the oldest stars, and the ages of the oldest star clusters are all less than 20 billion years, so this figure looks reasonable. Since the view from any galaxy is the same, each galaxy thinks that it was the center of the explosion. So there is no answer to the question, "Where was the center?"

236 / ASTRONOMY

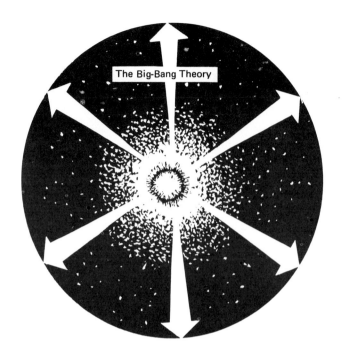

20-4 The universe started expanding from a tightly packed mass 20 billion years ago.

As time goes on, stars form from the gas clouds in all the galaxies. Eventually, when all their hydrogen has been converted to heavier elements, their supply of nuclear energy will be exhausted. The stars will die out; galaxies will go dim, then dark. The galaxies will be still more widely separated. The universe, according to the Big-Bang theory, will then be very different from what it is today.

This brings us back to the question of whether the universe has an edge. Whether or not it has something that we would recognize as a border, the Big-Bang theory implies that it has some sort of limit. Its volume keeps increasing as it expands, but it always has one definite total amount of matter. However many galaxies there may be, they all came from one single chunk of matter—the original sphere that exploded. That was all the matter there was and is. It exploded into something that became an expanding cloud of hydrogen gas. (Perhaps helium and some other elements were formed during the explosion, too.) This cloud broke up into separate clouds that later became galaxies. And each of them broke up into still smaller clouds that became stars, some with orbiting planets.

At some time, the formation of galaxy clouds stopped. According to the Big-Bang theory, there must be a definite limited number of galaxies. It does not matter how many. And all together, they have the same tremendous but limited (not infinite) mass as that original chunk of matter. So there is some kind of limit—even if we may never detect an edge.

Rugs on Curves

If a big sphere contains a certain number of cubic miles, you would expect it to have an edge, just like a big rug of as many square yards must have edges. Cosmologists, who study theories of the whole universe, get around this by using *curved space.* For instance, if your rug were big enough to cover the whole earth, it would have no edges because the earth's surface is curved. In the same way, cosmologists say, space is curved so that a light ray traveling in a straight line for several billion light years would come back to its starting point. The curvature is predicted by Einstein's theory to be caused by the density of matter (stars, gas, galaxies) in the universe. As the universe expands, the galaxies move apart, the density decreases, and there is more room—more cubic miles—but still no edge. In the rug-around-the-earth analogy (with square yards rather than cubic miles—two dimensions rather than three), the galaxies would be spots on the rug and expansion would be like the earth growing and stretching the rug, moving the spots apart.

This sounds logical. But is it necessarily right? If there is no center to the universe—if no galaxy has a preferred position, and it looks the same from wherever you view it—is it not possible that it looks the same *whenever* you view it? The *Steady-State* theory (shown in Figure 20-5), competing with the Big-Bang theory, does not suggest a sudden beginning 20 billion years ago. It suggests instead that the universe has always looked the same, and always will. It could, if new matter is being *created* at the rate of about one hydrogen atom per cubic mile per year. From that new

20–5 The universe is unchanging.

hydrogen, new galaxies can gradually be formed in the space between the older ones that are moving apart. The average distance between galaxies could always be about the same. The Steady-State theory assumes that new matter is being created all the time, instead of all at once as the Big-Bang theory assumes. The universe would then remain at the same density; there would always be young galaxies. In 100 billion years it would not be a starless universe as the Big-Bang theory predicts. Also, 20 billion years ago, the universe would have looked the same as it does now. And 20 billion years before that. In the Steady-State theory there is no beginning.

How can these rival theories be tested? According to the Big-Bang theory, the most distant galaxies should look like younger ones. They are farther away, so their light takes longer to reach us, and we see them as they looked longer ago when they were young. According to the Steady-State theory, the average age of galaxies is about the same whatever their distance; some are young, some are middle-aged and some are old. But in order to use these

THE UNIVERSE / 239

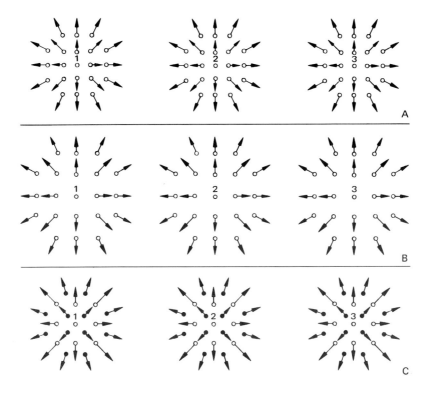

20-6
(A) The view of the universe from galaxy *1* (on the left) is identical to that from galaxies *2* and *3*—and from any other galaxy in the universe. Each circle represents a galaxy, each arrow a red-shift. (Notice that the arrows vary in length because the red-shift is larger at greater distances. Millions of galaxies should be pictured of course.)
(B) In the view from galaxies *1, 2,* and *3* five billion years later according to the Big-Bang theory, the retreating galaxies are more widely spread out. The view from all galaxies is still identical, but the over-all picture is different.
(C) According to the Steady-State theory, however, the view from galaxies *1, 2,* and *3* after 5 billion years will look just about like it did at first. This is because twelve new galaxies (shown by the solid circles) have formed.

two predictions to test the theories, astronomers must know how to tell a young galaxy from an old one. And as yet they don't.

Today, astronomers are arguing the merits of the two theories as vigorously as the Ptolemaic and Copernican theories were argued. They are searching for evidence, as Tycho and Galileo searched. In our lifetimes they may find it. Then will astronomy be finished?

New Mysteries, New Discoveries

There is no sign that astronomy will be finished. Certain recent discoveries have raised questions that nobody in Herschel's day or Hubble's day could have thought of, much less answered. For example, there is the puzzle of the *quasars*.

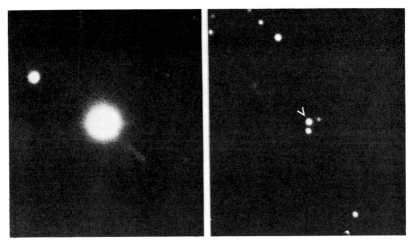

20-7 Quasars, like the ones pictured above, cannot be seen without a radio telescope. These new telescopes enabled astronomers to discover quasars and to gather much information about the universe.

As we have seen, radio "noise" comes from stars and planets. But in the 1950s some radio signals were found to be coming from areas of the sky where there are no stars visible. Finally, in 1963, Maarten Schmidt of Palomar Observatory found that these radio sources are in places on the celestial sphere where there are small bluish "star" images on telescope photographs. Are these images really stars? They look a bit fuzzy and Schmidt found that they have very large red shifts—showing speeds up to 100,000 miles per second. According to Hubble's Law, that would put them much farther away than any known galaxies. They had to be several billion light years away—much too far away to be stars. No single star could be seen at that distance.

So the quasars must be extremely luminous, both as radio sources and as light sources. To be that far away and yet to be visible, their luminosity must be over a hundred times that of the most luminous galaxies. Yet measurements of their apparent diameters showed that they are not giant galaxies. They are actually much smaller than average galaxies. No one knows of any nuclear reactions that could produce energy at such a high rate from such a small object.

These objects did not seem to be stars, but did not seem to be galaxies either. What to call them? Because the blue stars centered in the radio sources were so nearly points, they were called *quasi-stellar objects.* The name means starlike objects. Later, the name was condensed to *quasar.*

Quasars are really a puzzle. They are bundles of contradictions. Too far away and too fast-moving to be stars in our own Galaxy. Too small to be ordinary galaxies. Too bright to be so far away, or else too powerful (luminous) for any known way of producing energy.

Quasars would probably never have been noticed without the radio telescopes. They can only be seen or photographed through very large light telescopes. Without Hubble's Law, their distances would not be known. If they had been detected, they would have been taken for smaller, nearer masses of gas, within our Galaxy. If somehow Hubble's Law does not hold for quasars, then they may be much closer, not as luminous as we now think, smaller than we suppose, and with a much lower energy output.

However, astronomers have no reason to doubt Hubble's Law. Great distance and speed remain the only way to explain the quasars' enormous red shifts. So it is now considered that the quasars are very distant. It is possible that their high luminosities are due to violent explosions of oversized stars in galaxies of an odd type. Such explosions would be short-lived. As it happens, quasars are variable in brightness—possibly because their energy output depends on separate explosions of single stars.

Many explanations have been offered, but the problem remains unsettled. Arguments about them grow with the vigor of the old arguments: How was the solar system formed? What makes stars variable? Are faint oval clouds really galaxies? How was the universe itself formed—Big-Bang or Steady-State? Observations of quasars are occupying many astronomers today. New theories about them may have been suggested, or even perhaps proved, by the time you read this book.

One thing we can be sure of: the answer to the question of the quasars will only raise new questions. It will also lead to changes in our picture of the universe, as each new explanation has done in the past.

We, on the earth, are like passengers on a small spaceship in heavy traffic—revolving around our sun, which carries us with it in a slow path around the center of our Galaxy, which in turn is moving through the universe, who knows where, in the company of billions like it. The jewels that light the night sky are not ours alone—very likely we share them with many another observer on many another planet of many another "sun" in many another galaxy.

Test Yourself

1. What is revealed by a *red shift* in the spectrum of a star or distant galaxy?
2. Do any of the other galaxies show large blue shifts? What does this indicate?
3. The greater the red shift, the greater the what?
4. What simple but surprising idea is stated in Hubble's Law?
5. Does the fact that all observed galaxies are speeding away from the Milky Way Galaxy mean that our Galaxy is in the center of the universe?
6. How does the Big-Bang theory explain the expansion of the universe (flight of the galaxies away from each other)?
7. The Big-Bang theory states that as time goes by, the universe changes and so does its appearance to any observer inside it. But the Steady-State theory says the universe looks the same all the time, through all time. How can this be?
8. The Big-Bang theory offers an approximate age for the entire universe. How old is it supposed to be? What would the Steady-State theory say about the age of the universe?
9. According to the radio astronomers, radio signals from a moving source undergo something like a red shift or a blue shift. Yet a radio signal has no color. What is really happening to such a signal?

10. How were quasars originally discovered? Give a few reasons why the question of what they are has not been easy to settle.
11. How do astronomers know that the universe has been expanding for as long as several hundred million years?
12. Why does the Big-Bang theory suggest that the most distant galaxies should look the youngest?
13. Have astronomers decided how a young galaxy looks?

Index

aberration, of light, 91-93
absorption of energy, by planets, 130
absorption spectrum (dark-line spectrum), 131-135, 223
acceleration of gravity, 84, 88-89
Adams, John C., 118
Aldrin, Edwin, 154
Almagest, 7
Almanac (Kepler), 59
Alpha Centauri, 94
Andromeda spiral (M31), 221-222
Apollo missions, 154-156
apparent motion (*see* motion, apparent)
areas, equal, law of, 65
Aristarchus, 37
Aristotle, 49, 51, 55
arm, spiral Galactic, 215, 221-222, 223
Armstrong, Neil, 154
artificial satellites (*see* satellites, artificial)
astrology, 50
astronauts, 154-156
astronomical unit, 45, 63, 106
astronomy: new questions in, 240-242; observational, 50, 69-78, 103-105; radio, 181, 240-241; scope of, 1-2
atmosphere: of the earth, 150, 153; gravitational attraction and, 127-130; moon lacks, 150-154; of the planets, 127-130, 134-135, 136; and spectroscopy, 134-135
attraction, gravitational (*see* gravitational attraction)
axis of rotation, 15, 39

Baade, Walter, 215
Barnard's star, 199-200
barred spiral galaxy, 226, 227
Bessel, F. W., 94
Big-Bang Theory, 236-239
Big Dipper, 2, 11
black dwarf, 190
Bode, Johann, 115, 137
Bode's Law, 137-142
Bok, Bart J., 192
Bouvard, Alexis, 118
Bradley, James, 91-93
bright-line (emission) spectrum, 133-134, 222-223

brightness (magnitude) of stars, 170-171; *vs.* luminosity, 170

calorie, 128
Castor, 176-177
Cavendish, Henry, 105
celestial equator, 17-19
celestial objects, and natural law, 49-50, 51, 82; circular motion of, 35, 55
celestial pole, 13-14, 39
celestial sphere, 6-7; distance of, 8, 94-95; earth-centered theories, 8, 15; motionless, 37-46; rotation of, 11-19, 39; seasonal changes in, 22-25
center, Galactic, 208-211, 213, 215, 222, 223
center of mass, 89, 101-105, 177, 228
center of universe, theories of: earth at, 7-8, 15, 35, 37, 50; none exists, 235; sun at, 37, 42-46, 66
cepheids, 209; luminosity of, 225
Ceres, 138
chromosphere, 164-165
circular motion: *vs.* elliptical motion, 62-63; "perfection" of, 35, 55
climate, and erosion, 152, 153
cloud, dust (dark nebula), 192-194, 205-206
"clouds": catalogues of, 219-220; nature of, 220-223
cluster, galactic, 228-229
clusters, star: globular, 196-197, 209-210; H-R diagram of, 172-173; open, 195; seen as "clouds," 219-220; spectroscopy of, 223; and stellar evolution, 195-197
colors: spectrum of, 130-131, 161, 171-172; and temperature, 161, 171-175
coma (of comet), 144-145
comets, 51-55; composition of, 145; distance of, 52-55; evolution of, 145-147; light of, 143-147; and meteor showers, 142-147; orbits of, 55, 142-147; parts of, 115, 144-145
compounds, organic, 135-136
computers, in astronomical prediction, 109

244

constellations, 2–8; rotation of, 14; seen from earth, 8, 55–56; of zodiac, 16, 18, 32–33
continuous spectrum, 131
contraction: of nebulae, 193–194; of stars, 166–167, 187–188
conversion of matter to energy, 167, 186
Copernican system, 37–46, 49–67; demonstrated by Jupiter's satellites, 74–75; objections to, 49–52, 84; prediction in, 46, 59; *vs.* Ptolemaic system, 46, 56, 77–78
Copernicus, Nicholas, 37–46
core, galactic (*see* nucleus)
core, stellar, 187–188
corona, 164–165
craters, lunar, 153
creation of matter, 237–238
crescent moon, 27
crystalline sphere, 55
Curtis, H. D., 225
curvature of earth, 8–9, 31
curved motion, 87

daily motion: of celestial sphere, 11–15; of sun, 18
dark-line (absorption) spectrum, 131–135, 223
dark nebula (dust cloud), 192–194, 205–206
day, length of, 18–19, 41
degenerate matter, 189–190
Diaglogues Concerning the Two Principal Systems of the World, 78
Dipper, Big, 2, 11
direction of motion, 81–82, 88–89, 96–98
distance: and brightness, 170–171; and gravitational attraction, 89; parallax to measure, 52–56, 94–95, 169, 174; spectroscopy to measure, 206–207
Doppler, Christian, 178
Doppler shift: of galaxies, 233–234, 240; of stars, 177–178, 213
double planet, 149
double stars, 176–179
Dürer, Albrecht, 4
dust, interstellar: amount of, 207–208; clouds of, 192–194, 206; composition of, 208, spectroscopy of, 205–207
dwarf star: black, 190; white, 173, 184, 189–190

earth: age of, 235; atmosphere of, 150, 153; at center of universe, 8, 15, 35, 37, 50; climate of, and erosion, 152, 153; curvature of, 8–9, 31; distance of, from sun, 63, 106; eclipse by shadow of, 31; escape velocity of, 158; gravitational attraction of, 85–87, 158; mass of, 101–106; and moon, 71, 101–105, 149; orbit of, 38–39, 40–41, 63, 92–95; as planet, 38–41; rotation of, 37, 39, 96–99; wobbling of, 103–105
earthly objects, *vs.* celestial objects, 49–50, 82
earth-moon system, 101–105, 149
eccentricity, 63–64
eclipses: ecliptic and, 30–31; of moon, 29–31; prediction of, 31; of sun, 29–31, 164
ecliptic: and eclipses, 30–31; plane of, 18, 39–41; tilt of, 40
elliptical galaxy, 226, 227, 228
elliptical motion of planets, 62–63
emission (bright-line) spectrum, 133–134, 222–223
energy: matter converted to, 167, 186; of stars, 166–168, 183, 185–190, 194–197; of sun, 128–130, 166–168, 185
epicycles: in Copernican system, 46, 60, 64; in Ptolemaic system, 34–35, 37
equal areas, law of, 65
equator: celestial, 18–19; earth's, 19, 40–41; plane of the, 39–41
erosion, on moon, 152, 153
escape velocity: and atmosphere, 127–130, 150; and rocketry, 158, 160
evening star, 33
evolution, stellar (*see* stellar evolution)
expanding universe, 234
expansion of stars, 187–188

falling bodies, 84
First Account of the Revolutions of Nicholas Copernicus, 37
focus, 63
force: of gravity, 85–87; and motion, 81–82
form, perfect, circle as, 35
forward motion, 84, 87
Foucault, Jean, 96
Foucault pendulum, 96–98
Fraunhofer, Joseph, 131
Fraunhofer (dark) lines, 131–135, 223
freely falling bodies, 84
friction, and motion, 81–82
full moon, 27

galaxies: age of, 238–239; arms of, 221–222, 223; barred spiral, 226, 227; clusters of, 228–229; distance of, 220, 223–225, 231, 233–236; distribution of, 231–232; elliptical, 226,

227, 228; irregular, 226, 227; motion of, 228, 233-236; nature of, 220-222; normal spiral, 226, 227; nucleus (core) of, 222; number of, 225, 231-232, 237; observations of, 219-220; recession of, 233-234; rotation of, 228; spectroscopy of, 233; spiral, 221-222, 223, 226, 227; stars in, 223-225; types of 225-229. (*See also* Galaxy, the Milky Way)

Galaxy, the Milky Way: arms of, 215-216; center (nucleus) of, 208-211, 213, 222, 223; composition of, 205-208; mass of, 213; number of stars in, 203-204; origin of stars in, 203-204, 206; and other galaxies, 219-225; position of, in universe, 234-235; rotation of, 211-216; shape of, 204-206, 208-211, 213, 215, 222, 223; size of, 204, 208, 210-211; stars in, 73, 203-204, 206-207, 208, 209-210, 211-216; sun's position in, 208-211

Galileo Galilei, 69-78, 81-83, 149, 203

gases: in nebulae, 192; in planetary atmospheres, 127-130, 134-135, 136; spectroscopy of, 132-133, 175, 221

giant star: red, 173, 184, 187-188; supergiant, 173, 184

gibbous moon, 27

globular clusters: bubble of, in Galaxy, 210; luminosity of, 210; stars in, 196-197, 209-210

grains (sun's photosphere), 163

gravitational attraction: and acceleration, 84, 88-89; distance and, 89; and escape velocity, 127-130; force of, 85-87; mass and, 85-87, 89, 101-105, 108-109; and perturbation of orbits, 108-109, 117-118; universal law of, 89

Halley's comet, 142, 145
harmonic law, Kepler's, 65-66
heat energy of sun, received by planets, 128-130
heavens, perfection of, 35, 51, 55
helium, in stars, 186
helium reaction, 188
hemispheres, and seasons, 41
Henderson, Thomas, 94
Herschel, Caroline, 113, 176
Herschel, John, 220
Herschel, William, 111-116, 176, 203-204, 220
Hertzsprung, E., 172
Hertzsprung-Russel diagram, 172-175, 183-184

Hipparchus, 170
horizon, 5-7, 8
H-R diagram, 172-175, 183-184
Hubble, Edwin, 225, 231, 233
Hubble's law, 233-236, 240-241
hydrogen, in stars, 186
hydrogen-to-helium reaction, 167-168, 185-187, 194-197

inertia, 82, 84, 96-98
infrared, 131
inner planets, apparent motion of, 42
instruments, astronomical, 50
intensity, peak, 171
interstellar dust (*see* dust, interstellar)
invisible spectrum, 131, 180
irregular galaxy, 226, 227

Jupiter: observations of, 73-75; perturbation by, 108-109; red spot of, 126; satellites of, 73-75, 107

Kant, Immanuel, 219
Kepler, Johannes, 59-67
Kepler's Harmonic Law, 65-66
Kirchhoff, Gustav, 132-133, 222

law: of equal areas, 65; Hubble's, 233-237, 240-241; Kepler's Harmonic, 65-66; Newton's, of motion, 81-89; of planet motion, 64-67; of universal gravitation, 89
lead-ball experiment, 105
Leverrier, U., 119
life, extraterrestrial: conditions for, 135-136; probability of, 199
light: aberration of, 91-93; absorption of, 130; colors in (*see* colors); infrared and ultraviolet, 131; reddening of, by dust, 206-207; reflected, of planets, 28-29, 75, 128-130, 134-135, 140, 144; spectroscopy of (*see* spectroscopy); speed of, 169; wave motion of, 178. (*See also* luminosity)
light year, 169
lines, spectral: bright, 133-134, 222-223; dark, 131-135, 223
looping motion of outer planets, 33, 42-44, 46
Lowell, Percival, 19
luminosity: and brightness, 170; mass and, 179, 184, 185-186; stellar evolution and, 184, 185-186, 188; and temperature, 171-175, 184, 185-186
lunar eclipse, 29-31
Lunar Orbiter, 158-160

INDEX / 247

magnitude, 170-171
main-sequence star, 173-175, 179, 183-184, 185-187, 194-195
maps, sky, 2-8, 21, 22-25, 113, 116, 183, 219-220
Mariner missions, 159
Mars: absorption of energy by, 130; erosion on, 153; known to ancients, 33; orbit of, 59-63, 108-109; surface of, 125-126, 153; temperature of, 135
mass: center of, 89, 101-105, 177, 228; and gravitational attraction, 85-87, 89, 101-105, 108-109; and luminosity, 179, 184, 185-186; *vs.* weight, 101
matter: amount of, in universe, 237; conversion of, to energy, 167, 186; creation of, 237-238; degenerate, 189-190
Mercury: apparent motion of, 42; as morning and evening star, 33
meridian, 20-21
Messier, Charles, 219-220
Messier catalogue, 220
meteorite, 140-141
meteors: erosion by, 153; orbits of, 142-147; spectroscopy of, 141-142
meteor showers: comets and, 142-147; dates of, 143
Milky Way: appearance of, 203-204; stars in, 73. (*See also* Galaxy, the Milky Way)
minor planets, 138-142; discovery of, 138-139; mass of, 139; origin of, 139-140
missiles, 98-99
Mizar, 176
moon, the: age of 155; apparent motion of, 27-28, 30-31; astronauts on, 154-156; atmosphere lacking on, 150-154; composition of, 150, 154-155; crescent, 27; distance of, 29, 54-55; and earth, 71, 101-105, 149; and eclipses, 29-31; full, 27; gibbous, 27; light of, 28-29; mass of, 101-105; orbit of, 30-31, 103-105; phases of, 27-29; size of, 29; spectroscopy of, 150; surface of, 70-71, 149-154; temperature of, 150, 156; wobbling of, 103-105
moons, planetary (*see* satellites, planetary)
morning star, 33
motion: acceleration of, 84; change of, 81-82, 88-89, 96-98; curved, 87; direction of, 81-82, 88-89, 96-98, 213; Doppler shift to measure, 177-178, 213, 233; force and, 81-82; forward, 84, 87; Newton's laws of, 81-89; shared, 84, 98-99; sideways, of revolution, 101-105; wave, 178
motion, apparent: daily and yearly, 18; explanation of, in Copernican system, 56; motion of the observer and, 39, 52-53
motion, planet, laws of, 64-67
motion, proper, 213
Mount Wilson Observatory, 221, 231
M31 (Andromeda spiral), 221-222

nebulae: contraction of, 193-194; dark, 192-194, 206; dust in, 192-197; gas in, 192; gravitational attraction in, 193-197; spectroscopy of, 220
Neptune, discovery of, 117-120
New General Catalogue, 220
"new" star, 51. (*See also* nova)
Newton, Isaac, 83-89, 111, 130-131
Newton's laws of motion, 81-89
noon, 20
normal spiral galaxy, 226, 227
north celestial pole, 13-14, 39
nova: luminosity of, 223-224; and perfection of heavens, 51; supernova, 189, 224
nuclear reaction (*see* reaction, nuclear)
nucleus (galactic core), 208-211, 222, 223

observational astronomy, 50, 69-78, 103-105
observatories, 50, 221, 225, 231
observed motion (*see* motion, apparent)
observer: movement of, and apparent motion, 39, 52-53; position of, and relativistic viewpoint, 234
open cluster, 195
Orbiting Astronomical Observatory, 156
orbits: around center of mass, 101-105, 177; elliptical, 62-63; laws of motion of, 64-67; periods of, 42-44, 74; perturbation of, 108-109, 117-119; prediction of, 59, 109; wobbling of, 103-105, 177, 200
organic compounds, 135-136
outer planets, apparent motion of, 23, 42-44, 46
oxygen, and life, 136

pairs of stars, 176-179
Palomar telescope, 225, 231
parallax: of comets, 52-55; of stars, 55-56, 94-95, 169, 174
peak intensity, 171

perfection of the heavens, 51; and circular motion, 35, 55
period, orbital, 42-44, 74
perturbation, orbital, 108-109, 117-119
phases: of the moon, 27-29; of Venus, 75
photographic telescopy, 121-123, 138-139, 180, 181, 221
photosphere, 161-163
Piazzi, Giuseppe, 138
plane of the ecliptic, 18, 39-41; tilt of, 40
plane of the equator, 39-41
planets: apparent motion of, 32-36, 37-46; atmospheres of, 127-130, 134-135, 136; distance of, 44-45, 137-138; "double," 149; earth among, 38-41; energy of sun received by, 128-130; epicycles for, 34-35, 37, 46, 60, 64; formation of, 198-200; gravitational attraction of, 108-109, 127-130; inner and outer, 23, 42-44, 46; laws of motion of, 64-67; life on, 135-136, 200; mass of, 107-109, 115; minor (see minor planets); orbits of, 37-39, 40-41, 42-46, 55, 63-67, 82-83, 87-89, 108-109; period of revolution of, 42-44; on rings, 33-36, 55; rotation periods of, 126-127; satellites of (see satellites, planetary); spectroscopy of, 130-135; speed of, 46, 64-66, 88-89; table of data on, 123; temperatures of, 128-130, 135-136
Planet X, 117-119
Pluto, 120-123
Polaris, 2, 12
pole, celestial, 13-14, 39
Praesepe, 72
prediction: in astrology, 50; computers used in, 109; in Copernican system, 46, 59; of eclipses, 31; of orbits, 59, 109
pressure, radiation, 187
prism, 130-131
probe, space, 159
prominence, 163
proper motion, 212
Ptolemaic system: vs. Copernican system, 46, 56, 77-78; epicycles in, 34-35, 37
Ptolemy, Claudius, 8-9, 40, 41, 55-56
pulsation, star, 189

quasars, 240-242; distance, speed, and luminosity of, 240; size of, 241

radiation pressure, 187

radio astronomy, 181, 240-241
radio wave, 180
reaction, nuclear, 167-168, 185-190; helium reaction, 188; hydrogen-to-helium reaction, 167-168, 185-187, 194-197
recession of galaxies, 233-234
reddening, spectral, by dust, 206-207
red giant, 173, 184, 187-188
red shift, 233-234, 240
red spot of Jupiter, 126
red stars (of main sequence), 186
reflected light, planets shine by, 28-29, 75, 128-130, 134-135, 140, 144
reflecting telescope, 111-112
refracting telescope, 69, 111
rest, state of, 81-82
revolution (see orbits)
Riccioli, John, 176
rift, Galactic, 204-206
ring theory of planetary motion, 33-36, 55
rockets, 157-158
rotation: axis of, 15, 39; of celestial sphere, 11-19, 39; of earth, 37, 39, 96-99
Russell, Henry Norris, 173

satellites, artificial, 156-160; lifetime of, 158; orbiting of, 157-160; speed of, 158-160
satellites, planetary, 73-75, 149-156; and measuring mass of planets, 107, 115; period of, 74. (See also moon, the)
Saturn, 33
Schmidt, Maarten, 240
science, method, of, 77-78
seasons, 19-21; sky maps for, 22-25; sun and, 19, 20, 41; temperature of, 41-42
Shapley, Harlow, 209, 225
shared motion, 83, 98-99
shift, red, 233-234, 240
showers, meteor (see meteor showers)
sideways motion of revolution, 101-105
sky (see celestial sphere)
sky map, 2-8, 21, 22-25, 113, 116, 183, 219-220
skymark, 7
solar eclipse, 29-31, 163, 164
solar system, 46, 137-147, 149; Bode's law for, 137-142; formation of, 198-200. (See also Copernican system)
solar wind, 145
source, radio, 240-241
space probe, 158

spectrograph, 180
spectroscopy: atmosphere and, 134-135; distance determined by, 206-207; of gases, 132-133, 175, 220; motion determined by, 177-178, 213, 233; of sunlight, 130-131, 132, 134, 161; temperature determined by, 161, 171-172, 175
spectrum: colors in, 130-131, 161, 171-172; continuous, 131; Doppler shift of, 177-178, 213, 233-234, 240; invisible, 131, 180; lines in, 131-135, 220; photography of, 180; reddening of, by dust, 206-207
speed, uniform, 55, 64, 82
sphere, crystalline, 55. (*See also* celestial sphere)
spiral galactic arm, 214, 221-223
spiral galaxies: observations of, 221-222, 223; types of, 226, 227
spot, red, of Jupiter, 126
Sputnik, 159
"star," morning and evening, 33
star chart (*see* sky map)
Starry Messenger, The, 78
stars: age of, 238; brightness (magnitude) of, 170-171; clusters of (*see* clusters, star); color of, 171-175; composition of, 175, 179, 186, 189-190; in constellations, 2-8, 55-56; contraction of, 166, 187-188; core (nucleus) of, 187-188; distance of, 55-56, 94-95, 169, 170-171, 174, 206-207, 210; Doppler shift of, 177-178, 213; double, 176-179; dwarf, 173, 184, 189-190; energy output of, 167-168, 183, 185-190, 194-197; expansion of, 187-188; formation of, 191-200, 206, 215-216; in other galaxies, 223-225; in our Galaxy, 73, 203-204, 206-207, 208, 209-210, 211-215; gravitational attraction of, 187; in H-R diagram, 172-175, 183-184; light of, 15, 169-175, 178, 206-207; luminosity of, 170-175, 179, 184, 185-186, 188, 223-224, 225; main-sequence, 173-175, 179, 183-184, 185-187, 194-195; mass of, 177, 178-179, 184, 185-186; motion of, 14-15, 177-178, 211-216; "new," 51; nuclear reactions in, 167-168, 185-190, 194-197; number of, 203-204, 215; pairs of, 176-179; parallax of, 55-56, 94-95, 169, 174; planets of, 198-200; proper motion of, 212; pulsation of, 189; radiation pressure in, 187; red (of main sequence), 186; revolution of, in Milky Way Galaxy, 211-216; rising and setting of, 14-15; size of, 171, 184, 188; spectroscopy of, 171-172, 175, 177-178, 179-181, 206-207; speed of, 212-213, 214; stages of evolution of, 183-200, 203-204, 206, 215-216; telescopic observations of, 72-73; temperature of, 171-175, 184, 185-186, 188; types of, 51, 172-175, 179, 183-184, 185-188, 189-190, 194-195, 209-210, 223-224, 225; variable, 189, 209-210, 225. (*See also* sun, the)
Steady-State theory, 237-239
stellar evolution, 183-200, 203-204, 206, 215-216
stellar parallax, 55-56, 94-95, 169, 174
Struve, Wilhelm, 94
subsolar point, 128
sun, the: age of, 168; apparent motion of, 15-19, 30-31, 39, 40-41, 105; at center of universe, 37, 42-46, 66; color of, 161; contraction of, 166; distance of, 18, 106; eclipse of, 29-31, 163, 164; energy of, 128-130, 166-168, 185; evolution and future of, 168, 185, 197-198; gravitational attraction of, 87-89, 167; on the H-R diagram, 173-174; light of (*see* sunlight); mass of, 106; motion of, 38-39, 75, 105, 214; nuclear reactions in, 167-168, 185; parts of, 161-164; planetary motion around, 37-39, 40-41, 42-46, 55, 63-67, 82-83, 87-89, 108-109; position of, in Galaxy, 208-211; rotation of, 75; size of, 67, 106, 163; solar wind, 145; spectroscopy of, 130-131, 132, 134, 161; telescopic observation of, 75; temperature of, 161, 166
sunlight: angle of, and seasons, 19, 20, 41; reflected by planets, 28-29, 75, 128-130, 134-135, 140, 144; spectroscopy of, 130-131, 132, 134, 161; stars in, 15
sunspots, 75, 161-163; cycle of, 163
supergiant, 173, 184
supernova, 189, 224

tail, comet, 115, 144-145
telescope: allowing for aberration, 92-93; motor driven, 138-139; photographic, 121-123, 138-139, 180, 181, 221; radio, 181, 240-241; reflecting, 111-112; refracting, 69, 111
temperature: and atmospheres of planets, 128-130; color and, 161, 171-175; and life, 135-136;

luminosity and, 171–175, 184, 185–186; of seasons, 41–42; spectroscopy to determine, 161, 171–172, 175
time exposures, 120–121, 221
Titius, Johann, 137
Titius' progression (*see* Bode's law)
Tombaugh, Clyde, 121
twilight, 150
Two New Sciences, 78
Tycho Brahe, 49–58, 59

ultraviolet, 131
uniform speed, 55, 64, 82
universal gravitation, law of, 89
Universe: age of, 197, 236–239; center of, 8, 15, 35, 37, 42–46, 50, 234–237; edge of, 231–232, 237–239; expansion of, 234; our Galaxy in, 234–235; mass of, 237; motion in, 232–236; size of, 225, 231–232
Uranus, 114–116; discovery of, 115; distance of, 138; mass of, 115; orbit of, 115, 117–119

van de Kamp, Peter, 200
variable star, 189, 209–210, 225
Vega, 94
velocity, escape (*see* escape velocity)

Venus: absorption of energy by, 130; apparent motion of, 42; as morning and evening star, 33; telescopic observation of, 75

waning of moon, 27, 28
water, and life, 135–136
wave: light, 178; radio, 180
waxing of moon, 27, 28
weighing the planets, 101–109
weight, *vs.* mass, 101
Whipple, Fred L., 145
white dwarf, 173, 184, 189–190
Wilson Observatory, 221, 231
wind, solar, 145
wobbling, orbital: of Barnard's star, 200; of earth and moon, 103–105; of star pairs, 177; of sun, 105
Wright, Thomas, 203, 219

X, Planet, 117–119
x-ray, 180

yearly motion of sun, 18
Yerkes Observatory, 221

zeneith, 2, 6–7
zodiac: planets in, 32–33; sun in, 16–18